T0313350

Security within CONASENSE Paragon

RIVER PUBLISHERS SERIES IN COMMUNICATIONS

Series Editors

ABBAS JAMALIPOUR
The University of Sydney
Australia

JUNSHAN ZHANG
Arizona State University
USA

MARINA RUGGIERI
University of Rome Tor Vergata
Italy

Indexing: All books published in this series are submitted to the Web of Science Book Citation Index (BkCI), to SCOPUS, to CrossRef and to Google Scholar for evaluation and indexing.

The "River Publishers Series in Communications" is a series of comprehensive academic and professional books which focus on communication and network systems. Topics range from the theory and use of systems involving all terminals, computers, and information processors to wired and wireless networks and network layouts, protocols, architectures, and implementations. Also covered are developments stemming from new market demands in systems, products, and technologies such as personal communications services, multimedia systems, enterprise networks, and optical communications.

The series includes research monographs, edited volumes, handbooks and textbooks, providing professionals, researchers, educators, and advanced students in the field with an invaluable insight into the latest research and developments.

For a list of other books in this series, visit www.riverpublishers.com

Security within
CONASENSE Paragon

Editors

Ramjee Prasad

CTIF Global Capsule (CGC)
Aarhus University
Denmark

and

Global ICT Standardization Forum for India (GISFI)

Leo P. Ligthart

CONASENSE
The Netherlands

LONDON AND NEW YORK

Published 2019 by River Publishers

River Publishers

Alsbjergvej 10, 9260 Gistrup, Denmark

www.riverpublishers.com

Distributed exclusively by Routledge

4 Park Square, Milton Park, Abingdon, Oxon OX14 4RN

605 Third Avenue, New York, NY 10017, USA

Security within CONASENSE Paragon / by Ramjee Prasad, Leo P. Ligthart.

Routledge is an imprint of the Taylor & Francis Group, an informa business

ISBN 978-87-7022-092-7 (print)

While every effort is made to provide dependable information, the publisher, authors, and editors cannot be held responsible for any errors or omissions.

Contents

8   The Network Neutrality in Service Innovation Era   105

Yapeng Wang and Ramjee Prasad

Preface

Integration of Communication, Navigation, Sensing and Services (CONASENSE) is one of the leading research areas below the sunshade of CTIF Global Capsule (CGC). CGC is a non-profit organization that has a goal of strengthening all academic research in the field of technology, its applications in various sectors and then convergence with business strategies to strive the smart global ecosystem. CGC sights to guide and oversee its aspirants to explore new horizons in the field of Scientific research.

The present book, Security Within CONASENSE Paragon is the 7th volume in the series of CONASENSE and consist of 11 chapters.

Chapter 1 throws light on the cyber threats and security aspects of CONASENSE with details and emphasizes the need for Cyber Security for the future ICT ecosystems and Cyber-Physical systems. The heterogeneous architecture of CONASENSE poses various privacy and security issues. Almost all the existing and upcoming cyber-physical systems are dependent on Information and Communication Technology Infrastructure (ICT).

Chapter 2 focuses on the effective use of data aggregation techniques and applied network security from CONASENSE perspectives. The chapter identifies the potential issues and provides the remedy to overcome. It may take different forms according to the functionality of protocol stack used for sensing, communication and Navigation of data with appropriate resource utilization in the network.

Chapter 3 tells about the Reliable and Secure Low Power Wide Area Networks (LPWANs) communication in Interference Environments. It is depicted that the establishment of a large variety of LPWANs is of the utmost importance. The realization of LPWAN applications faces several challenges with significant problems of LPWANs like interference, and the considerable attenuation of signals in the industrial, scientific and medical (ISM) range has been described in detail.

Chapter 4 describes the integration of the two-way communication network, smart grid components, and the data they generate is a demonstration of the Communications, Navigation, Sensing, and Services (CONASENSE) model. The need for effective cloud security analytics is becoming more evident due to cyber-attacks. The smart grid infrastructure can be protected using data analytics, specifically security analytics. The security analytics tools collect the 'big' data generated by the smart grid segments and then analyze it to detect and proactively alert on potential cyber-attacks.

Chapter 5 presents the Crowd density estimation in public spaces can be a highly effective tool for establishing global situational awareness. Cost-effective solutions for these systems should be not invasive and mainly, privacy preserving. The chapter makes an overview of the most exciting business applications as well as an overview of traditional and more novel approaches. The chapter also discusses the opportunity to bring location/context-based services, and the associated substantial market opportunities, as protagonists in 5G systems.

Chapter 6 introduces the individual Unmanned Aerial Vehicle (UAV) nodes in a network that can transmit data to each other. UAV communication networks are usually called Flying Ad Hoc NETworks (FANETs). FANETs are still in the development stage and face many challenges. The chapter identifies the challenges to see which technologies have been put forward and examine whether those technologies can overcome the FANET challenges and what possible consequences they may have.

Chapter 7 explains some conventional and recent technical directions towards overcoming the challenges in developing robust and accurate methods of moving object detection. In many computer vision applications, robust and real-time foreground detection is a crucial issue. It is found that this approach is normally useful for a stationary camera but does not work well for rotating camera. The challenge, in this case, is to segregate the dynamic entities while the background turns virtually at a fixed pace.

Chapter 8 reviews the development of Network Neutrality (NN) debate process, and the opinions from different sides, including the network providers, the service providers, other relevant companies, governments and researchers. This chapter also introduces a telecommunication convergence concept that is CONASENSE (Communication, Navigation, Sensing and Services). It aims to formulate a vision on solving societal problems with new telecom technique to improve human welfare benefit. This chapter focuses on the service of CONASENSE and summarizes the current situation of NN in the service innovation era.

Chapter 9 demonstrates that digitalization is a natural product of achieving ubiquitous connectivity, enabling seamless digital experience through all-around availability of various data, which in turn is a driver for business model innovation (BMI). This chapter also addresses two research questions. Firstly, it explores digitalization as business value co-creation catalyst. To this end, three case studies are presented. Secondly, it examines the security challenges and opportunities arising from rapid digitalization. Finally, it offers an approach to enabling security in the digital ecosystem through BMI.

Chapter 10 provides a review of the several countermeasures that have been devised, especially against the spoofing threat, both at the data and signal level and discuss their effectiveness concerning the context and the possible side information available to the attacker and receiver. Finally, we will propose novel models for the analysis and design of partial spreading code encryption and the integration of side information coming from other sensors into the integrity verification from the computed GNSS trajectory.

Chapter 11 illustrates Future Integration of Communications, Navigation and Sensing technologies (CONASENSE) are challenged by finding profitable business models (BM). 5G technologies are costly to develop and implement, and the prices for buying 5G licenses are high and expected to become even higher. The focus of the chapter is to unwrap future CONASENSE BM's and BMES's required to "carry" the future CONASENSE technologies. The chapter also discusses how BMI in a world of future CONASENSE technologies and related BMES's is expected to be carried out and implemented. Future Integration of Communications, Navigation and Sensing technologies (CONASENSE) are challenged by finding profitable business models (BM). 5G technologies are costly to develop and implement, and the prices for buying 5G licenses are high and expected to become even higher.

We hope that readers will enjoy reading the book and learn in depth the role of security within CONASENSE paragon.

List of Contributors

Albena Mihovska, *Department of Business Development and Technology, Aarhus University, Denmark; E-mail: amihovska@btech.au.dk*

Dimitriya Mihaylova, *Technical University of Sofia, Bulgaria; E-mail: dam@tu-sofia.bg*

Dnyaneshwar S. Mantri, *Sinhgad Institute of Technology, Lonavala, Pune, India; E-mail: dsmantri@gmail.com*

Ernestina Cianca, *Department of Electronic Engineering, University of Rome Tor Vergata, Italy; E-mail: cianca@ing.uniroma2.it*

Georgi Iliev, *Technical University of Sofia, Bulgaria; E-mail: gli@tu-sofia.bg*

Homayoun Nikookar, *Netherlands Defence Academy, The Netherlands; E-mail: h.nikookar@mindef.nl*

Mauro De Sanctis, *Department of Electronic Engineering, University of Rome Tor Vergata, Italy; E-mail: mauro.de.sanctis@uniroma2.it*

Nicola Laurenti, *Department of Information Engineering, University of Padova, 35131 Padova, Italy; E-mail: nil@dei.unipd.it*

Peter Lindgren, *Aarhus University, Denmark; E-mail: peterli@btech.au.dk*

Ramjee Prasad, *Arhus University, Arhus, Denmark; E-mail: ramjee@btech.au.dk*

Sarmistha De Dutta, *CTIF Global Capsule, Aarhus University, Herning, Denmark; E-mail: sd56@columbia.edu*

Shivprasad P. Patil, *NBN Sinhgad School of Engineering, Pune, India; E-mail: shivprasad.patil@sinhgad.edu*

Silvia Ceccato, *Department of Information Engineering, University of Padova, 35131 Padova, Italy; E-mail: ceccatos@dei.unipd.it*

Simone Di Domenico, *Department of Electronic Engineering, University of Rome Tor Vergata, Italy; E-mail: simone.didomenico@uniroma2.it*

Sofoklis A. Kyriazakos, *Department of Business Development and Technology, Aarhus University, Denmark; E-mail: sofoklis@btech.au.dk*

Tommaso Rossi, *Department of Electronic Engineering, University of Rome Tor Vergata, Italy; E-mail: tommaso.rossi@uniroma2.it*

Vandana Milind Rohokale, *Sinhgad Institute of Technology and Science, Pune, India; E-mail: vmr.301075@gmail.com*

Viktor Stoynov, *Technical University of Sofia, Bulgaria; E-mail: vstoynov@tu-sofia.bg*

Vladimir Poulkov, *Technical University of Sofia, Bulgaria; E-mail: vkp@tu-sofia.bg*

Yapeng Wang, *International Cooperation Department, China Academy of Information and Communications Technology, MIIT, China; E-mail: wangyapeng@caict.ac.cn*

Zlatka Valkova-Jarvis, *Technical University of Sofia, Bulgaria; E-mail: zvv@tu-sofia.bg*

List of Figures

List of Tables

List of Abbreviations

AI	Artificial Intelligence
AUH	Aarhus University Hospital
BIAS	Broadband Internet Access Services
BM	Business Model
BMES	Business Model EcoSystem
BMI	Business Model Innovation
BMIL	Business Model Innovation Leadership
BS	Background Subtraction
BSs	Base Stations
CAGR	Compound Annual Growth Rate
CDMA	Code Division Multiple Access
CGC	CTIF Global Capsule
CNN	Convolutional Neural Network
CONASENSE	Communication, Navigation, Sensing and Services
CPs	Content Providers
CV	Computer Vision
D2D	Device-to-Device
D2H	Device-to-Human
D2M	Device-to-Machine
Diffserv	Differentiated Service
DL	Deep Learning
DNN	Deep Neural Network
ECG	Electrocardiogram
eNBs	enhanced NodeBs
FCC	Federal Communication Committee
FDMA	Frequency Division Multiple Access
FG	Foreground
FTC	the Federal Trade Commission
FWN	Future Wireless Network
GAN	Generative Adversarial Network
GMM	Gaussian Mixture Model
GNSS	Global Navigation Satellite System
GSM	Global System for Mobile Communications

HMM	Hidden Markov Model
ICT	Information and Communication Technology
IMU	Inertial Measurement Unit
IntServ	Integrated Services
IOD	Issue of Data
IoT	Internet of Things
ISPs	Internet Services Providers
ITS	Intelligent Transportation Systems
LBS	Location-Based Service
LOS	Line of Sight
LTE	Long Term Evolution
M2M	Machine-to-Machine
MAC	Message Authentication Code
MBMI	Multi Business Model Innovation
MEO	Medium Earth Orbit
ML	Machine Learning
NB	No Blocking
NBBM	Network Based Business Model
NFV	Network Functions Virtualization
NLOS	Non Line of Sight
NN	Network Neutrality
NPP	No Paid Prioritization
NT	No Throttling
PRN	Pseudorandom Code
PVT	Position, Velocity and Time
QoL	Quality of Life
QoS	Quality of Service
RAN	Radio Access Network
RFID	Radio-Frequency Identification
RGB	Red Green Blue
RNN	Recurrent Neural Network
ROI	Return on Investment
RSVP	Resource Reservation Protocol
SCER	Secret Code Estimation and Replay
SDN	Software-Defined Networking
SLAM	Simultaneous Location and Mapping
TCO	Total Cost of Ownership
TESLA	Timed Efficient Stream Loss-tolerant Authentication
UAV	Unmanned Aerial Vehicle
UMTS	Universal Mobile Telecommunication System

1

Cyber Security-The Essence of CONASENSE

Vandana Milind Rohokale[1] and Ramjee Prasad[2]

[1]Sinhgad Institute of Technology and Science, Pune, India
[2]Arhus University, Arhus, Denmark
E-mail: vmr.301075@gmail.com; ramjee@btech.au.dk

CONASENSE is the essential integration of communication, navigation or localization and sensing capabilities for the upcoming 5G wireless communication. The heterogeneous architecture of CONASENSE poses various privacy and security issues. Almost all the existing and upcoming cyber-physical systems are dependent on Information and Communication Technology Infrastructure (ICT). On one side, the dependence on ICT reduces manual intervention and errors, increases the speed of operation but on the other side safety, security and efficiency is compromised. Lawful regulatory framework for cyber standards, security and privacy is presently in its baby stage. This chapter throws light on the cyber threats and security aspects of CONASENSE with details and emphasizes the need of Cyber Security for the future ICT ecosystems and Cyber-Physical systems.

1.1 Introduction

The term CONASENSE refers to the seamless integration of services like communication, navigation and positioning, and sensing for making provision of new services. CONASENSE is a technical society which inspires for research among all these integrated domains. The world is looking forward to welcome the fifth generation (5G) mobile communication which is visualized as a convergence of existing wired, wireless and upcoming next generation technologies with embedded intelligence. Worldwide users demand ultra-fast services with more efficiency in terms of less resource consumption and user centric approaches. Whenever integration of some services is visualized, first thing to be given utmost importance is privacy and security [1].

Upcoming future generation wireless systems demand convergence of technologies with extremely high speed and capacity, minimum possible delays, very high energy and cost efficiency with high reliability, privacy, security and good QoS and QoE for users. Communication started with tools like drums, horns, heliograph, newspapers, postal letters, telegraph, telephone, radio, computer, television, computer networks, Internet, Mobile communication has now reached to smart phones with various capabilities embedded in them. Various technologies supporting communication are cellular communication, Wi-Fi, Wi-MAX, wireless sensor networks (WSN), Internet of things (IoT), visible light communication (VLC), cognitive radio networks (CRN), cooperative multiuser MIMO, satellite communication, aviation communication technology, optical fiber communication, etc. Smart communication systems make use of navigation and positioning techniques to take intelligent decisions for command and control the situations. Cloud computing services are very much essential for providing of virtual memory spaces. For the upcoming converged next generation wireless networks, heterogeneity is the main basic aspect. The conceptual visualized structure of CONASENSE is depicted in Figure 1.1.

These interfaces are the integral part of the next generation wireless networks. Spectrum sensing for cognitive radio networks and the converged scenario is the more challenging task. The research for upcoming technologies has taken a very good pace. But with this progress, the hackers and white collar people with their malicious activities are equally making progress in finding out innovative ways to break the available security provisions. Really users are looking for user centric services with usable security and privacy preserving technologies in the next generation wireless networks. The chapter is organized as below. Section II throws light on the cyber-attacks and threats related to CONASENSE. Section III elaborates possible security threats to CONASENSE. Section IV deals with the cyber security standards for some cyber systems related to CONASENSE. Section V discusses the general framework for cyber security provision in the CONASENSE. Section VI concludes with hugefuture research scope in this area.

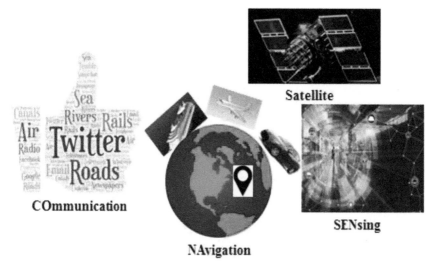

Figure 1.1   CONASENSE conceptual structure.

1.2 Known Cyber Attacks Related to CONASENSE

On 1st September 1983, the Soviet Union shot down a Korean Airlines flight 007 causing death of all 269 passengers on board. Since the flight was originated in New York, the US President at that time, decided to allow the civilian use of global positioning system (GPS). The usage of GPS was limited only to military applications due to the security and safety reasons [2]. GPS and global navigation satellite system (GNSS) proved to be blessings for communication services because they provide location and time information in all weather conditions. But the same things were utilized as weapons by the attackers to penetrate the aircraft information systems for the malignant purposes.

Unmanned Aerial Vehicles are the remote controlled or fully automated aircrafts without any manual human intervention on board. UAVs are used for variety of applications such as surveillance, transportation, armed attacks in military services, online purchase delivery, etc. These vehicles are required to collect and process large amount of data. Because of the huge amount of data and sensitive kind of information enclosed by UAVs make them very fascinating object for cyber-attacks. UAVs are prone to theft, data manipulation, spying, GPS spoofing, etc. Iranian Military UAV RQ-170 Sentinel, US UAV fleet, MQ-9 Reaper, and AR Drone are some of the eye opening examples of the cyber-attacks on the UAVs [3].

Table 1.1   Some known cyber attacks on CONASENSE related cyber systems

Sr. No.	Compromised Cyber System	Attack Description	Source
1.	A Boeing production plant in Charleston, South Carolina	Due to WannaCry ransomwear cyberattack, an US based Boeing production plant was attacked by shutting down the whole production system.	[5]
2.	Kiev's Airport and Metro System	PETYA Ransomware attack targeted and compromised Ukranian Kiev's Airport and Metro system by halting the services for users.	[6]
3.	Australian Government and Corporate Computer Networks	The Australia Trade Commission, Austrade and the Australian defense research division suffered due to cyber-attacks on the extremely confidential plan details about a Geostationary Communication Satellite.	[7]
4.	Industrial Controls of a German Steel Mill	Cyber-attack penetrated the industrial controls of a German Steel Mill and caused huge damage by keeping a blast furnace continuously on and keeping it away from getting shut down.	[8]
5.	South Korean GPS System	Global Positioning System (GPS) and Global Navigation Satellite System (GNSS) signal jamming attack lasting nearly a week and affecting signal reception of more than 1000 aircrafts and 700 ships.	[9]
6.	Airlines and Airports	Recently US, Turkey, Spain, Sweden and Poland aircrafts have been plagued with security breaches like introduction of malware causing delays, information loss and always tensed feeling of getting tracked.	[10]
7.	Unmanned Aerial Vehicle (UAV)	GPS Spoofing Attack- US UAV RQ-170 Sentinel is supposed to be hijacked by Iranian Forces by compromising GPS Satellite signaling system.	[11, 12]
8.	Korean Airlines Flight 007	The Korean flight which started from New York for Seoul was compromised in the Soviet Airspace and it was shot down which resulted in the death of all 269 on board people.	[13]

Airplane entertainment systems have been utilized to provoke cyber-attacks in the plane. Ground control systems, air navigation service, and many communication links connecting them can face severe cyberattacks like hacking, jamming, spoofing, gaining remote access to the aircraft cockpit system, gaining control and remotely changing the flight programs. Many a times GPS system of the aircraft has been endangered to intentional and unintentional targeting and disruption by cyber attackers form the country or abroad hackers [4].

Industrial control systems may be the next victims of cyber-attacks. For manufacturing, transportation, energy, oil, gas, and chemical industries, monitoring and control of their industrial and physical infrastructure has become very important because of the cyber threats and attacks. Some known cyber-attacks which terribly experienced different cyber systems are listed in Table 1.1.

1.3 Cyber Threats to CONASENSE

The wireless world is looking forward to see CONASENSE as a complete platform for upcoming wireless communication era. It is the essential integration of various communication technologies, navigation techniques for air, road, and sea travel, sensing services necessary for satellite communication and other services. Because of multiple communication and other service interfaces, CONASENSE platform has to face lot of security threats. The attacks can be primarily mentioned like Availability, Integrity, and Confidentiality as depicted in Figure 1.2.

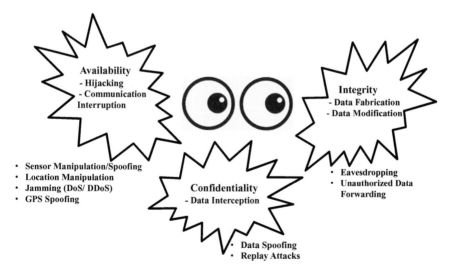

Figure 1.2 Probable cyber attacks on communication, navigation and sensing services.

Availability attack mainly relates to hijacking and communication interruption attacks. It may further consist of sensor manipulation or spoofing attack, location manipulation attack which can prevent the correct data collection from various entities. Jamming attack under availability threat may deny transmitting the genuine data to the network or receiving legitimate data from the network. GPS spoofing attacks may intentionally change the travel path.

Integrity attack is related to unauthorized data fabrication and modification in the original data or information. Instead of authentic data, fake data can be sent. Replay attack can resend or reuse the sensitive data without proper permission. Data capture and interception can happen under confidentiality attack. Eavesdropping may cause leakage of confidential information. The trustworthy information can be further retransmitted to the unauthorized malicious users. These types of attacks can be performed for monetary gains or malign purposes. They can further add delays in the emergency services. The success of some emergency monitoring systems is largely dependent on timely availability of the sensitive data. It may further result into the failure of such critical missions.

1.4 Cyber Security Standards Related to CONASENSE

Cyber security standards are the set of rules laid down for the smooth and safe functioning of the cyber systems. These standards contain policies, rules and regulations, security concepts, guidelines, risk assessment techniques, training practices, various actions, safety assurance, etc. Cyber security standards are necessary for the cyber systems to ensure risk reduction, prevention and mitigation of cyberattacks, etc. These cyber systems include users, networks, hardware, softwares, various processes, information at different stages of data transmission, applications, services, and systems associated with wired and wireless networks.

European Telecommunication Standards Institute (ETSI) has formed technical committee for cyber security named as TC CYBER in 2014 [14]. ISO/IEC 27001 to 27006 are the standards for information security management system (ISMS). ISO/IEC 27002 standard is a high level guide to cyber security laid down by International Organization for Standardization (ISO) and the International Electrochemical Commission (IEC) in 2013. North American Electric Reliability Corporation (NERC) and National Institute of Standards and Technology (NIST) are some of the standards

Table 1.2   Cyber security standards for cyber systems related to CONASENSE

Sr. No.	Cyber System	Security Standards	Reference
1.	Industrial Control System (ICS) Environment.	ANSI/IS A-62443-3-2	[15]
2.	Guide to Industrial Control Systems (ICS) Security for ICS/SCADA Systems.	NIST SP800-82r2	[15]
3.	For improvement in Resilience Act OR Cyber Air Act to address the importance of aircraft security including guidelines about development of cybersecurity for aviation industry and timely reporting of cyber-attacks to Governments.	Aircraft Cybersecurity Standards – Cyber Air Act (Proposed)	[16]
4.	Information Technology Security Techniques – Guidelines for Cybersecurity.	ISO/IEC 27032:2012	[17]
5.	Wireless Systems for Industrial Automation: Process Control and related applications.	ANSI/IS A-100.11a-2011	[18]

which are modified with cyber security provisions. Table 1.2 shows the list of some cyber security standards for some CONASENSE related cyber systems.

1.5 CONASENSE Cyber Security Framework

It is the need of time to develop the robust and usable cyber security framework for the communication, navigation and sensing related services. Initially, there is need to set up common cyber standards for this integrated CONASENSE system. Cyber security culture should be created and penetrated among all the layers of society and services. The users and industrial systems should be intelligent enough to quickly recognize the risks, threats, their intents and the actors behind particular malign action. If the attack happens, then the users should openly speak about it. The much required incident response should be made available in time. Strong defensive systems should be developed which can withstand such attacks.

Sensible development of design principles and operational principle strategies is necessary to combat such attacks. Innovative and curious research efforts are necessary which can continuously work on various cyber security issues. Academia should encourage research scholars to undertake such research projects which will be beneficial for the society. The government, industry and academia should walk hand in hand for this cause. Society related issues should come from industry to academia as research problems.

Figure 1.3　General framework for CONASENSE cybersecurity.

Academia should develop manpower to supervise such research projects by encouraging their research community. Government should keep sufficient funds for these kinds of crucial research projects and research themes. The general framework for development of cyber security for such integrated concept like CONASENSE is depicted in Figure 1.3. It shows eleven steps are the essential stepping stones for the development of cyber security system standards.

1.6 Conclusions and Future Scope

For the integrated system like CONASENSE, privacy, security and trust are the major concerns. Next generation wireless communication users expect higher data rates, lesser delays, reliable, efficient and secure communication with low cost. This chapter throws light on the cyber threats and attacks related to the CONASENSE related systems. There is huge scope for the researchers to work on usable security for such system which will provide strong yet cost efficient security solutions to the users. Communication services are suffering from various threats like intercepting the information from cellular communication networks, Wi-Fi Networks, etc.

Aviation systems like commercial airplanes, civil aircrafts and UAVs are in need of reliable and secure navigation and positioning systems so that users can trust them. With the upcoming Internet of Things (IoT) into reality, the attention needs to be given to the security aspects of wireless sensor networks, cognitive radio networks, and other interfaces of fifth generation wireless communication. For the heterogeneous environments, the interfaces among various networks demand efficient and secure spectrum sensing and spectrum sharing. Public safety and security is of prime importance while designing and developing such kind of integrated systems. Geo sensor network (GSN) and Machine to machine (M2M) communication via satellites are the next big things to work on cyber security research.

References

[1] Ernestina Cianca, Mauro De Sanctis, Albena Mihovska and Ramjee Prasad, CONASENSE: Vision, Motivation and Scope, Journal of Communication, Navigation, Sensing and Services, Vol. 1, 1–22, 2014.

[2] Sarah Laskow, —How a 1983 Plane Crash Helped Speed Up Civilian Use of GPS‖, Nextgov- The Atlantic November 3, 2014. http://www.nextgov.com/big-data/2014/11/how-1983-plane-crash-help ed-speed-civilian-use-gps/98000/

[3] Kim Hartmann, ChristophSteup, —The Vulnerability of UAVs to Cyber Attacks – An Approach to the Risk Assessment‖, 5th International Conference on Cyber Conflict K. Podins, J. Stinissen, M. Maybaum (Eds.), NATO CCD COE Publications, Tallinn, pp. 1–23, 2013.

[4] Deepika Jeyakodi, Air and Space Law, Leiden University, —Cyber Security in Civil Aviation‖, For EALA Prize, July 2015.

[5] Nick Statt, "Boeing production plant hit with WannaCryRansomware attack", THE VERGE, March 2018. https://www.theverge.com/2018/3/28/17174540/boeing-wannacry-ransomware-attack-production-plant-charleston-south-carolina

[6] Olivia Solon, Alex Hern, "'Petya' ransomware attack: what is it and how can it be stopped?", The Guardian, June 2018. https://www.theguardian.com/technology/2017/jun/27/petya-ransomware-cyber-attack-who-what-why-how

[7] Linton Besser, Jake Sturmer and Ben Sveen, Four Corner ABC-News, —Government computer networks breached in cyberattacks as experts warn of espionage threat‖, 29 Aug 2016. http://www.abc.net.au/news/2016-08-29/chinese-hackers-behind-defence-austrade-security-breaches/7790166

[8] USAFEATURESMEDIA, cyber war news, ‖Cyber-attacks have already targeted critical infrastructure around the world and the U.S. could be next‖, December 11, 2015. http://www.cyberwar.news/2015-10-09-cyber-attacks-have-already-targeted-critical-infrastructure-around-the-world-and-the-u-s-could-be-next.html

[9] Reuters, The Maritime Executive, —South Korea Revives GPS Backup after Cyber Attack, May 2016. http://www.maritime-executive.com/article/south-korea-revives-gps-backup-after-cyber-attack

[10] Jorge Valero, EurActiv.com, —Hackers bombard aviation sector with over 1,000 attacks per month, June 2016. https://www.euractiv.com/section/justice-home-affairs/news/hackers-bombard-aviation-sector-with-more-than-1000-attacks-per-month/

[11] CNN Wire Staff, —Obama says U.S. has asked Iran to return drone aircraft, 2011.

[12] Todd Humphreys, —Statement on the vulnerability of civil unmanned aerial vehicles and other systems to civil gps spoofing‖, Austin, 2012.

[13] Alyssa Newcomb, NBC News, Tech-Security, "Hacked in Space: Are Satellites the Next Cybersecurity Battleground?" October 2016. http://www.nbcnews.com/tech/security/hacked-space-are-satellites-next-cybersecurity-battleground-n658231

[14] https://portal.etsi.org/webapp/WorkProgram/Report_WorkItem.a sp?WK I_ID=45906

[15] Kathy Trahan, tripwire, —The State of Security, ICS Security, Industrial Control Systems: Next Frontier for Cyber Attacks? June 2016. https://www.tripwire.com/state-of-security/featured/ics-next-frontier-for-cyber-attacks/

[16] Josephine Wolff, Future Tense, The citizen's guide to the future, ‖Hacking Airplanes‖, May 2016. http://www.slate.com/articles/technology/future_tense/2016/05/the_aviation_industry_is_starting_to_grapple_with_cybersecurity.html

[17] Maria Lazarte, ISO, ‖Are you safe online? New ISO standard for cybersecurity‖, October 2012. http://www.iso.org/iso/news.htm?refid=Ref1667

[18] ISA Standards, InTech Magazine, ‖ISA-100 wireless standard receives resounding approval in IEC technical voting Nov/Dec2013. https://www.isa.org/standards-and-publications/isa-publications/intech-magazine/2013/december/standards-isa-100-wireless-standard-receives-resounding-approval-in-iec-technical-voting/

2

Layered Network Security for Efficient Data Aggregation in CONASENSE

Dnyaneshwar S. Mantri[1] and Ramjee Prasad[2]

[1]Sinhgad Institute of Technology, Lonavala, Pune, India
[2]Department of Business Development Technology, Aarhus University,
Herning, Denmark
E-mail: dsmantri@gmail.com; ramjee@btech.au.dk

In today's technological era information is the key to success. It is essential that information be secured so that it is only available to those authorized to access it in correct and squashed way. The data collection and security are prime concern in the big data handling networks. The compression of data from multiple, heterogeneous sensor along with security can no longer be a set-and-forget solution, but instead needs to be effectively developed, thoughtfully implemented, and continuously managed process. It provides not only the privacy, authentication in handling the data communicated from lower layer to upper layer but also increase the efficiency. Also, the energy required for communication and navigation is more as compared to sensing and computation. In this context the chapter focuses on the effective use of data aggregation techniques along with layered network security from Communication, Navigation, Sensing and Services (CONASENSE) perspectives. The chapter identifies potential issues in design of secure network and application of aggregation methods with remedy to overcome. The layered protocol stack used for sensing, communication and navigation of data uses hybrid model (Security + Aggregation) for resource utilization and to improve the quality of service parameters.

Security within CONASENSE Paragon, 11–34.

2.1 Introduction

Recent advances in CONASENSE provides more flexibility in daily life applications with the processes of sensing, communication and navigation of data from sources to sink. It can be used with reference to the different range of applications used in Wireless Sensor Network (WSN) such as e-agriculture, habitat monitoring, vehicular technology, health-care, military, smart homes and environmental monitoring and control. In all these use-cases, each node has specific task of sensing the event and send data to sink. However, nodes used in the development of sensing and communication network are resource constrained and have limited processing, storage, and communicating capabilities [1–3]. The data processing capacity of the WSNs depends on the number of nodes used in the formation of the network. The generated data is huge and need the pre-processing as aggregation to reduce the multiple data, by minimizing the redundancy. In this context the energy consumption, network lifetime and communication bandwidth of the WSNs is greatly influenced by the changes in the network topology and the way of aggregating the data [4, 5]. The process of sensing, communication and nevigation with data handling prospective is affeted by performance metrics used by the Physical, Data link and Network layers of WSN protocol stack. Therefore, a need arise to develop a cross layer abstraction for improving the network performance with metrics as energy consumption, network lifetime, delay, bandwidth utilization, and computational overheads in WSNs.

One of the effective way is to add the security in the netwok and perform the aggregation on the multiple copies of data to reduce the redundancy. The data aggregation process may use different approaches as,

In-network Aggregation: Suitable for localized events and only critical information is passed to sink.

Grid based Aggregation: Adapts dynamic changes in topology suitable for mobile environment. It reduces the traffic.

Tree-based Aggregation: Suitable for In-network processing, Spanning Tree Mechanism (SPT) is the effective way to improve the performance parameters of the network.

Cluster-based Aggregation: Effective in all types of scenario and node types. Network is scalable and stable by overcoming the communication overheads. The grouping of nodes in the cluster and CHs in the network improves the energy consumption [6, 7]

The cluster-based aggregation is preferred over flat since clustering improves the scalability by stabilizing the network topology. In WSN, the network is operational even if one Cluster Head (CH) fails. It involves a lower delay of communication since nodes perform short range transmissions to the CH and have simple routing structure. The application of additive and divisible aggregation function (min, max, avg, sum, count, median, etc.) at CH results in the reduction of number of packets to be transmitted from the source node to sink. It saves the energy, improves the network lifetime and bandwidth utilization [5].

In the hierarchical WSN, resource allocation is related to the amount of bandwidth given to the CH used for communication of aggregated data with link between source node to sink by one hop or Multi-hop distance. While communicating the information from source to destination data may be added by the malicious node behaving as router and changes the original contents, since each node in the network has a specific task of sensing the environmental conditions such as temperature, humidity or pressure at the particular time. Under static conditions, the data packets transmitted by nodes are aggregated at CH. Also, each CH sends several copies of aggregated data directly to sink which increases energy consumption in communication rather than sensing and presents the fundamental limit on network lifetime of WSNs. The maximizing network lifetime with data gathering considers the optimal tree, where CH in route may act as an aggregator or simply forward the data. The current research trend in cluster-based data aggregation is to minimize the trade-off between energy, delay and throughput at the sink in correlation with added security.

The chapter has been organized in different sections as; Section 2.2 describes the essential requirements of security by focusing on communication model with functionalities and issues in utilization of WSN protocol stack. Section 2.3 gives in-depth analysis of data aggregation models, performance metrics and describes the aggregation methods. Section 2.4 explores the different security models used along with threats in security and aggregation. Finally, Security, Threat, and Aggregation model is presented along with conclusions in Section 2.5.

2.2 Essential Requirements of Layered Network Security

Sensor network consists of numerous, heterogeneous nodes distributed either in random, quasi-random, or deterministic way. They form infrastructure-less architecture to sense the event of interest and collect the data according to the applications. With increased node density in the network, an independent communication of sensed data to the base station or sink causes congestion. The excessive traffic at sink node may increase the response time and loss of packet and data. This is caused due to resource constraint nature of the node such as low bandwidth, low memory, and small battery. This directly effects on the performance of the network in terms of energy consumption and network lifetime. Also, improper utilization of link bandwidth causes the congestion and loss of data. Each layer in the TCP/IP model has specific task and need to be performed without any addition of data from malicious nodes while transferring from one layer to another. The security and threat management is the important consideration. Security deals with the Confidentiality (encryption), Integrity (identity management, digital signatures), and Availability (protection from denial of services). Security is an important architectural component of sensor networks. The attacks may happen at the sources or at the data collection centre. With the ubiquity of sensors in the network, security has been a matter of grave. The communication architecture along with possible location of threats is shown in Figure 2.1.

The possibility of attack on the data is either at source or aggregator. To avoid data loss and maintain confidentiality during user interface, the network security entails securing data against attacks while it is in transit.

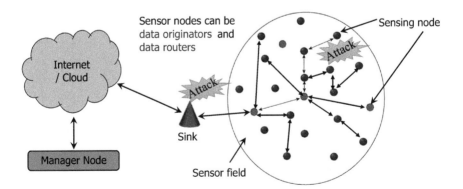

Figure 2.1   Communication architecture of network.

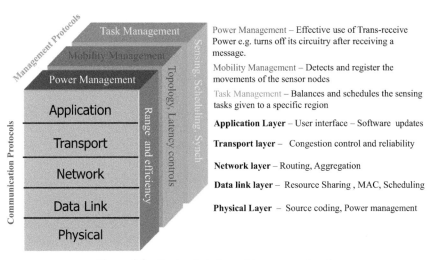

Figure 2.2   Protocol stack used in sensor network.

2.2.1 Protocol Stack used by Sensor Network

The protocol stack used by the sink and all sensor nodes is shown in Figure 2.2. This protocol stack combines power and routing awareness, integrates data with networking protocols, communicates power efficiently through the wireless medium, and promotes cooperative efforts of sensor nodes [1]. It has two planes of abstraction for transfer of data from application layer to physical layer along with consideration of Task, Mobility and Power management.

1. Communication Protocols

It decides the performance metrics and sets the conditions used for communication, navigation and services used by the network. In one of three domain sensor node spends more energy in communication rather than sensing and navigation. The protocol stack used by sensor network has following layers.

a. Physical layer: Minimization of energy consumption of radio is of paramount importance while designing the physical layer for WSN in addition to the usual effects such as scattering, shadowing, reflection, diffraction, multipath, and fading.

b. Data link layer: The data link layer is responsible for the multiplexing of data stream, data frame detection, medium access and error control. Ensures reliable point-to-point and point-to-multipoint connections in a communication network.

MAC protocol for sensor network must have built-in power conservation mechanisms, and strategies for the proper management of node mobility or failure: Code, Time and Frequency are the measure parameters for the design. Data link layer decides the resource sharing mechanisms by managing the schedules of node.

c. **Network layer:** This layer is important and decides the transfer of packets from transport layer to lower layer. Secure routing decides and identifies the best possible route. The basic issues and functionality of network layer when designing for the WSN are:

   o Power efficiency – Depends on the effective selection of route from source to destination.
   o Data centric – The nature of the data (interest requests and advertisement of sensed data) determines the traffic flow.
   o Data aggregation – is useful to manage the potential implosion of traffic because of the data centric routing rather than conventional node addresses. An ideal sensor network uses attribute-based addressing.
   o Location systems, i.e. ability for the nodes to establish position information.

   Also, the inter-networking with external networks via gateway or proxy nodes in the network layer is considered with

   o Datagrams – Connection less network-does not need to define predefined path.
   o Virtual circuit – Connection oriented networks – need to define the path for packet switching.

d. **Transport layer:** It demands for the end-to-end reliability in transfer of data. In reality, the TCP variants developed for the traditional wireless networks are not suitable for WSNs due to: Multiple senders, re-tasking and reprogramming for same event. The main advantage of transport layer is, it minimize number of transmissions to recover lost information and operate correctly even in situations where the quality of the wireless link is very poor. Some of the silent features of transport layer are

   o Self-configuration – Dynamic topology of network with addition of mobility.

○ Energy awareness – Sensor nodes are notified to decrease their frequency of reporting if the reliability level at the sink node are above the minimum.

○ Congestion control – Takes advantage of the high level of correlation between the data flows corresponding to the same event.

○ Collective identification – Sink only interested in the collective information from a group of sensors not in their individual reports.

e. Application Layer: The little work of ingredient has been reported on meticulous area, for e.g. turning sensor radio on-off, Authentication, key distribution for security purpose, possible attribute distribution for sensor nodes, and advertisement of sensed data etc. Some potential application protocols such as sensor management protocol (SMP), task assignment, data advertisement protocol (TADAP), and sensor query and data dissemination protocol (SQDDP), have been suggested.

2. Management Protocols:
These protocols are used to minimize the power consumption by coordinating and assigning the sensing task for each node.

a. Power Management: It considers the trans-receiver power used by the hardware of node in communication of data. The radio model used by the node will be turn off when message is received or transmitted, Also, if any node is running with low power it communicates to nearest node regarding not participation of data in routing which helps to save the power and can be utilized for sensing only.

b. Mobility Management: With heterogeneous nature of nodes in the network and addition of mobility, network dynamics frequently changes with need to detect the new route for communication of message by establishing link with nearest neighbour. In all, mobility management plan detects and registers the movement of node for reduced power consumption along with route back.

c. Task Management: It is most important in accordance with saving the power by balancing the schedules and sensing tasks given to the node.

All details of Protocol stack used in communication, Navigation, Sensing, and Services (CONASENSE) are summarised in the Table 2.1.

Table 2.1   Protocol stack functionality and issues

Layers	Functionality	Issues
Upper layer (Application)	In-network applications: Application Processing, Data aggregation, Query processing and External data bases	Queuing and modelling, Time dependant arrivals
Layer 4 (Transport)	Data dissipation and accumulation, Catching and Storage	Congestion detection and avoidance, Scalable and robust communication
Layer3 (Network)	Topology management and Routing	Malicious attacks, Interference, and Penetrations
Layer2 (Data link)	Contention, Channel sharing(MAC), Timing and Locality	Collision avoidance, Channel sharing, Data compression
Layer1 (Physical)	Communication channel, Sensing, Actuation, and Signal processing	Connectivity of sensors, Bandwidth allocation, Selection of channel.

2.2.2 Secure Network Design Issues in CONASENSE

Traditionally, network design has followed a layered communication architecture in which each layer of protocol stack handles the specific network functions. By providing standardize interface, layered architecture provide a high degree of modularity and interoperability among heterogeneous networks. The layered communication network generally has trade-off for energy efficiency and delay due to transparency between layers for modularity and optimization in network behaviour. Some of the basic issues and challenges for design of CONASENSE network are explored in [8]. The design issues in CONASENSE with the secure network using sensors are shown in Figure 2.3.

Network Dynamics: Static, Heterogeneous or Mobile (Flat or Clustered Network). Frequent Changes in the network architecture demands more resources to be utilized.

Node Deployments: Random, Quasi-random or Deterministic – Continuous update of routing tables for efficient route finding.

Energy Consumption: Depends on the transceiver power, distance (single, multiple Hop), traffic, route selection etc.

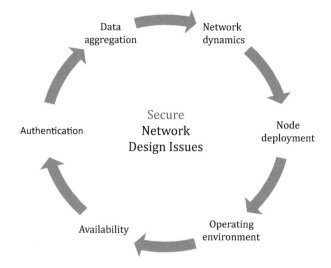

Figure 2.3 Network design issues.

1. Reliable transport (encryption),
2. Bandwidth and power limited transmission
3. Data centric routing,
4. In-network processing,
5. Self-configuration,
6. Operating environment
7. Hardware constraints – Transmission media, Radio-frequency integrated circuits, Power constraints, Communication network interface, etc.
8. Network architecture and protocols, Topology and fault tolerance, Scalability, Self-organization and Mobility.

Operating Environment: Noisy or application specific, requires sophisticated approaches to integrate information across temporal and spatial scales.

Availability: It gives the guarantee for the survival of services provided by network against Denial-of-Service (DoS) attacks. In the view of security, privacy and data aggregation availability of CH as aggregator is more important than source nodes used for sensing and communication of information to CH.

Authentication: It is important to provide the Integrity check on the data changed by the malicious nodes. It provides the identification of node participating in the network. For message exchange between two nodes authentication can be provided by use of symmetric key cryptography.

Data Aggregation: Reduces the communication overheads, delay and energy consumption. It also guarantees the message integrity, authentication and confidentiality during transfer of aggregated packets though different layers of protocol stack. In-network processing helps to reduce the effect of redundant data generated by nodes.

In nutshell,

- Wireless Sensor nodes have the unique capability of transmission at different power levels. With variation in transmission power, a trade-off exists between a number of hops and overall communication bandwidth available to individual nodes.
- Real-time applications require guarantees in terms of throughput and end-to-end delay.
- As traffic introduced into the network is in the form of events, may cause congestion and leads to in-efficient resource utilization [energy and bandwidth].
- Enabling of Medium Access Control (MAC), Network and Transport layer can provide energy efficient behaviour in the network. This emphasizes on the cross – layer design with fine-grained optimization in the network based on interlayer cooperation.
- According to the evolution of technology from 1G to 5G and usage in the CONASENSE it is necessary to provide the secure communication between layers with increased demand of energy saving, low utilization of communication bandwidth with reduced delay.

2.3 Secure Data Aggregations in CONASENSE

Data aggregation techniques are used to solve the problem of implosion in WSNs with added security. The main purpose of CONASENSE is to have effctive gathering of data sensed by sensors, decide the medium of communication, get effective routing from source to destination without any addition of malicious data during transit. This problem arises when packets carrying the same information arrive at node and sensed by more than one node. This is different than the problem of "duplicate packets" in conventional adhoc networks. Here it is the high level interpretation of the data in the packets is

Table 2.2 Requirements of secure aggregation protocols

Papers	Data Confidentiality	Encryption Method	Data Integrity	Source Authentication
PDA [9]	Yes	End-to–end Symmetry	Yes	Yes
DyDPA [11]	Yes	End-to–end Symmetry	Yes	Yes
Ozdemir et al. [12]	–	End-to-end Symmetry	Yes	Yes
L.Zhu et al. [13]	Yes	Cryptographic Hop-by hop	Yes	Yes

what determines if the packets are the same. Even the information sensed by nodes may take different form. The sensed information will be aggargated by use of min/max aggregation function into a single packet. In this regard data aggregation can be considered as data fusion. The data communication senario with possible location of attacks is shown in Figure 2.1. Data coming from multiple sensor nodes are aggregated, if they have about the same attributes of the phenomenon being sensed, then they reach a common routing or relaying node on their way to the sink. In this view the routing mechanism in a sensor network can be considered as a form of reverse multicast tree.

The security protocols used in the data aggregation considers the End-to-end aggregation, Hop-by-hop aggregation and two levels of aggregation security. Table 2.2 summarizies characteristics and requiremnts of secure data aggregation protocols.

2.3.1 Data Aggregation Model

The information sensed by the node is specific and periodic. Each node generates the multiple copies of data and send to the CH. The information generated by the source node is represented by its entropy, i.e. average information and is given by Equation (2.1)

$$H_n(d) = H_1 + (n-1)\left[1 - \left(\frac{1}{\frac{d}{c}+1}\right)\right]H_1 \qquad (2.1)$$

Where H_1 – Entropy of single source, n – Number of nodes, d – Inter node spacing, and c – Correlation level.

The information collection may be at the source nodes from the interest of events or at the sink. As the traffic introduced in the network is uneven and information generation rate is different, under such a senario the rate-based aggregation could serve the purpose. The rate–based aggregation functions applied at the source node and sink for data accumulation consider the spatial and temporal correlation of packets.

If X_i and Y_j are two variables representing correlation between number of packets generated by the *'u'* and *'h'* participating nodes in the cluster, provided that $i = 1. \ldots$ K and $j = 1 \ldots$ M, then

1. The perfectly compressible aggregation function is given by Equation (2.2) [3]

$$f(A) = \sum_{i=1}^{K}(X_i) + \frac{1}{M}\sum_{j=1}^{M}(Y_j) \qquad (2.2)$$

For $\forall(X_i) = $ different rates and $\forall(Y_j) = $ equal rate
2. Cost of aggregation is given by Equation (2.3)

$$f(C_A) = \sum_{j=1}^{h}(E_j) \quad h = \% \text{ of heterogeneous nodes} \qquad (2.3)$$

Data from 's' nodes is compressed sequentially before routing to the sink. The Energy costs as the function of cluster size is obtained from Equation (2.4)

$$Ec(c) = nH1\left[1 + \frac{(S-1)}{2(1+c)} + \frac{D}{S} + \frac{(S-1)D}{(S)(1+c)}\right] \qquad (2.4)$$

Where n – Number of nodes, H_1 – Enetopy, S – Cluster Size, c – Correlation, D – Spacing between node.

2.3.2 Open Issues and Performance Metrics in Data Aggregation

Recent advances in WSNs provide more flexibility towards different range of applications such as e-agriculture, habitat monitoring, vehicular technology, health care, smart homes, environmental monitoring and control. In all these applications, each node has a primary task of sensing the event and communicate data to the sink. However, sensor nodes used in the WSN is resource constrained and has limited processing, storage and communicating capabilities. The open issues and performance measurement factors that greatly effects on the development of application specific WSN used in CONASENSE are

- Fault tolerance – The failure of nodes should not severely degrade the overall performance of the network
- Scalability – The mechanism employed should be able to adapt to a wide range of network sizes (number of nodes)
- Cost – The cost of a single node should be kept very low
- Power consumption – Should be kept minimum to extend the useful life of network.
- Hardware and software constraints – Sensors, location finding system, antenna, power amplifier, modulation, coding, CPU, RAM, operating system must be operative with minimum cost.
- Topology maintenance – In particular to cope to expected high rate of node failure.
- Deployment – Pre-deployment mechanisms and plans for node replacement and/or maintenance
- Environment – Static and mobile conditions with Noisy or noise less.
- Transmission media – ISM bands, infrared, etc.
- Clock Synchronization – The synchronization algorithms used introduces the time drift causing overheads and errors. It consumes significant energy due to miss-match of clock. Also, packets gets collide and need retransmission causing more bandwidth requirements. Synchronization algorithm need to reduce the overheads and clock drifts and utilize the available energy and bandwidth efficiently.
- Mode of communication (single or multi-hop) – To increase the throughput and reduce the energy consumption one-hop communication is preferred but consumes more bandwidth within the cluster. During network-wide communication, it is better to use the multi-hop, since direct transfer of data from node to sink will require more energy.

The performance parameters for effective utilization of network in terms of measured value gives the impact of security on network layers. These parameters are affected by the addition of malicious nodes in the network which produces the dummy packets. Some of the metrics used by layered network for performance measures are,

- **Throughput:** It is a measure of the aggregated information at the CH or sink. It depends on deployment strategies, the aggregation methods, and allocation of scheduling slot for transfer of data. It has direct impact on the link bandwidth.
- **Energy Efficiency:** It depends on the amount of data gathering and the functionality of the nodes used in aggregation and retransmission of packets. It also depends on the failure of the relay nodes in the path

Table 2.3 Node and network level operational challenges

Node Level	Network Level
• Energy constraints	• Energy-efficiency at all layers
• Limited storage and computations	• Data aggregation
• Scalability	• Scheduling
• Clock synchronization	• Time synchronization
• Low bandwidth	• Node placements (localization)
• High error rates	• Network scalability
• Privacy and security	• Self- organized routing
	• Data dissemination
	• Data protection

from source to sink. It consumes more energy in route calculations with variation in network topology.

- **Latency:** Delay is one of the prime concerns in data aggregation mechanism. It depend on the one-hop or multi-hop communication, number of collisions in channel and depth of aggregation tree.

- **Bandwidth:** Nodes used have very low bandwidth and cannot outfit for real-time applications. If the nodes in WSN generates busty-traffic, it demand more bandwidth for enhanced performance of the network. The bandwidth utilization mechanisms need to be developed according to the variable traffic (packet generation rate) from the nodes. We can correlate the throughput with bandwidth utilization– collision avoidance.

- **Network Lifetime:** It depends on the how much energy is drained in the transmission of broad cast messages for formation of cluster, election of CH, data gathering and scheduling activities of nodes. It also depends on the nodes contending for channel accesses.

- **Security:** Adding security to aggregated data is a critical task. It needs to increase the usability of the channel with minimum attacks.

- **Mobility and heterogeneity:** The mobility of node or sink frequently changes the network dynamics causing increased energy and bandwidth requirements for finding the optimal path. But it has benefit of increasing the throughput with reduced delay. By adding the nodes with different energy network sustains more and increases the lifetime.

The network and node level challenges are given in Table 2.3.

2.3.3 Data Aggregation Methods

Nodes used in the sensing and collection of information regarding events of interest are scarce in the resources as bandwidth, energy and storage capacity.

If communication of data packets is direct, it may causes the loss of data due to flooding at the CH or sink, less availability of required communication bandwidth, the collision of packets, change of traffic patterns, decision on data processing and finding optimal route. The reliable communication of data to CH or sink demands more bandwidth which is scarce in WSNs [1, 2]. The current bandwidth management mechanisms focus on the physical layer parameters to optimizing its utilization. The modulation and coding schemes gives solution for increasing the throughput and energy saving in the wireless networks. However, theses mechanisms may not fulfil the requirements of WSN, since nodes used for formation of network are resource constrained. As WSNs are event-based and generate the multiple data, it necessitates the need for effective and secure data aggregation protocol strategy which will provide the appropriate solution for communication and navigation of aggregated data with proper utilization of resources (Bandwidth, Energy, and Storage) for sustaining network.

Figure 2.4 shows the different approaches used for effective collection of data with bandwidth utilization according to the network dynamics. The data aggregation protocols used for information gathering and security essential can be classified according to the structure used for the formation of the network, some protocols use Cluster/ Tree based approach with in-network

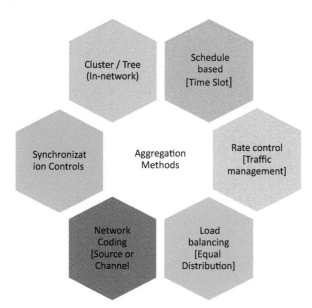

Figure 2.4   Data aggregation mechanisms.

processing, Time slotted, Rate control, Load balancing and Network coding techniques for effective and secure data aggregations. The approach used in all the methods may be of data centric routing, data aggregation with low level naming, aggregation with path sharing, optimal or adaptive aggregation, aggregation based on determination point or multilevel fusion [2].

Data Aggregations: The basic purpose of data aggregation algorithms is to reduce redundant data form nodes with in-network processing. It helps to reduce the communication cost and saves the energy. The factors affecting on the data aggregation algorithms are; network topology, formation of cluster, election of CH, node deployment, heterogeneity of nodes, mobility patterns, traffic patterns, transmission media, link quality and aggregation methods. The in-network processing and data management has trade-off between computation and communication complexity [2–5]. The classification of heterogeneous data aggregation security protocols and trust controls are explored in [9–14].

Slotted Time Approach: TDMA or fusion of TDMA and CSMA, in this approach guaranteed time slot is considered but faces the problem of bandwidth wastage, if the node does not have data to transmit. It is not suited for the multi-hop communication. The proper allocation of slots for transfer of data improves the energy consumption and bandwidth utilization [15].

Schedule-Based Approach: Making node radio on-off according to the availability of sensed data improves the performance of network. It faces the problem of adjusting sleep time and synchronizing the slots. The metrics used for improving network performance are, buffer management, duty cycle adjustment for sleep and wakeup time of node, management of data flows according to availability of channel, collision avoidance etc. Scheduling fails when the data available to transmit is high and wastes the bandwidth [15].

Synchronization Controls: Un-synchronized network increase the overheads and errors in matching the global clock of network with local clock of node. It consumes more energy and performance degrades. Spanning tree-based synchronization helps to reduce the synchronization errors occurred due to clock drifts and increases the throughput [16].

Rate Control Approach: It is considered according to the traffic introduced in the network. If real-time traffic is introduced, it must restrict to its packet transmission rate according to the allocated share. The rate control can be done at various layers of the protocol stack. A network with increased data and fixed bandwidth causes the congestion and has a direct impact on energy consumption packet service time, and throughput. The fair sharing of bandwidth is decided according to variation in network traffic and congestion control mechanisms. The congestion control technique reduces the energy consumption and improves the channel quality [17].

Load Balancing Approach: It is based on the distribution of the number of nodes in each cluster with same characteristics. The configuration and equal-sized cluster is crucial for extending the network lifetime. Even distribution of nodes leverages the data delay [18].

Network Coding: This approach is used to encode the data for reliable communication of packets. Channel coding theorems reduces the error probability with an increase in the throughput. Distributed source coding is one of the major concerns [19]. Redundant data from nodes is one of the reason of consuming more energy and bandwidth. Source coding is one of the way to reduce the redundancies in the sensed data while channel coding increases the reliability. The way of compression of data is important.

These schemes may be incompatible with some of techniques that rely on promiscuous packet reception to improve the network performance. The bandwidth and energy constraint of node has direct impact on QoS and efficiency of WSN used in CONASENSE. The communication bandwidth utilization is affected by the performance metrics used by the physical, data link and network layers of WSN protocol stack. Therefore, a need arise to develop a cross layer abstraction for improving the network performance with metrics as energy consumption, network lifetime, delay, bandwidth utilization, and computational overheads in WSNs.

2.4 Security Models in CONASENSE

The data aggregation techniques such as rate-based approach, load balancing at upper layers and so on demand the security in order to improve the functionality of layers. Very little work has been addressed by use of cross layer approach (MAC + Network) with focus on aggregation, and security concerns of nodes and network under mobile and heterogeneous scenarios. This section explores the different security models and controls used in CONASESNE for possible reduction of attacks, improvement in communication bandwidth and energy consumption.

2.4.1 Layered Network Model

The layered security model used in the CONASENSE along with possible data controls, security checks and protection is shown in Figure 2.5.

At the initial stage of network security model, data from user is accessed according to the application and passed to the next layer. The data obtained may be static or dynamic. The data has security and protection checks in order to reduce the redundancy added by malicious node. It is also equally important to have control on topology due to movement of nodes in the network. If topology changes are frequent, then it is difficult to maintain the routing table and ultimately affects on the energy consumption and utilization of bandwidth due to one-hop or multi-hop communication. To have effective implementation of security and aggregate the information followings points are more important.

1. Protection of data and validation checks
2. Providing access to authenticate users
3. Minimum topological variations
4. System level checks
5. Control over access networks
6. Identity management

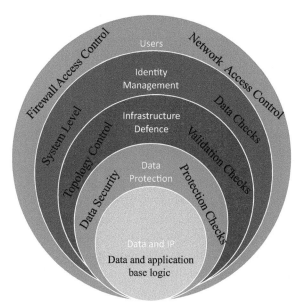

Figure 2.5 Layered network security model.

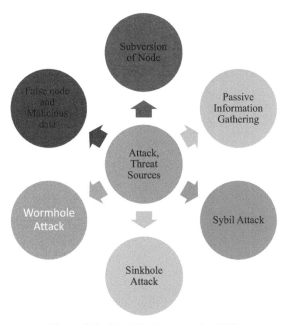

Figure 2.6 Possible threats in the LNS.

2.4.2 Threats in Layered Network Security (LNS)

The possible threats in layered network security with aggregation scenario at node and sink are shown in Figure 2.6. The possible threats in the LNS are, Subversion of Nodes with same identity, Passive information gathering by malicious node in the process of aggregation, Inclusion of false route, Sybil attack, Sinkhole attack and Wormhole attack.

The malicious insider threat is one that can cause the greatest damage, as the insider is a trusted individual with access to systems containing sensitive data and in-depth knowledge of the weaknesses in those systems.

2.4.3 Layered Security Controls

Numerous security controls used in designing a multi-layered security infrastructure considers the preventive and detective controls for efficient utilization of resources for getting required security tasks. The details regarding the preventive and detective controls applied at the different protocol layers of WSNs are summarized in Table 2.4.

Table 2.4   Preventive and detective controls in layered security [20]

Protocol Layers	Preventive Controls	Detective Control
Physical Layer	Security, Policies and Procedure	Security Training, Policies and Procedure
Network Layer	Network Access Control, Encryption, Malware detection and prevention	Security configuration management, Change control, Incident altering
Host Layer	Malware detection/ prevention, System updates, User access controls	Security Configuration Mgt. Log monitoring, Susceptibility management
Application Layer	Software patching and updates, System hardening	Change control, Log monitoring, Susceptibility management
Data Layers	User access controls, Encryption	File integrity monitoring, change control, Log monitoring

The Preventive Security Controls must take following points

1. High prevalence to malwares by updated software
2. Removal of unnecessary services
3. Setting of access rights to correct users
4. Authenticated users are allowed to access the operational network
5. Protection against threats, invalid data occurrence.
6. Proper coding and encryption of received aggregated data

While Detective Security Controls should follow following points

1. Variations in the data could be detected and exploited as threat.
2. Log monitoring and compare of previous data with current.
3. Regular test of security controls
4. Responding to the suspected security incidence in a timely manner.

2.4.4 Security, Attack and Aggregation Model

The integration of Security, Attack and Aggregation model used in CONASENSE is shown in Figure 2.7. The CONASENSE technology imposes some risk concerns, such as weak protection to security and privacy at different levels. The auditing process to this technology becomes challengeable due to distant and remote connectivity. The essential requirements of the security are Data Availability, Data Freshness, Data Confidentiality, Data Integrity and Source Authentication. The possible attacks on these

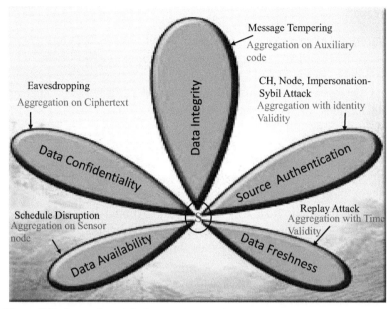

Figure 2.7   Integral module for security, attack and aggregation in CONASENSE.

are Schedule disruption, Replay attack, Eavesdropper, Message tempering and Node Impersonation (Sybil Attack) respectively. The integration of data aggregations at the source with security plays the important role in design of network used for CONASENSE.

2.5 Conclusions

The layered network security helps to mitigate the problems of redundant data addition by malicious node and improves the quality of service parameters. The data aggregation technique with specially correlated grouping of nodes in the lower layer of network and clustered grouping at upper layer reduces the complexity in analyzing the data or packets generated, reduces the communication overheads and energy consumption by nodes and cluster heads. Data aggregation along with the security provides good level of accuracy and resolutions in CONASENSE paradigm. It has the integration of all enabling technology used in future access network with good sense of utilization. Future work can be extended with cloud computing as base for storage and analysis of aggregated data.

References

[1] I.F. Akyildiz, W. Su, Y. Sankarasubramaniam, and E. Cayirci, "Wireless sensor networks: a survey," Journal of Computer Networks, 38(4), pp. 393–422, March 2002.

[2] Fasolo E, Rossi M., Widmer J, and Zorzi M. "In-network aggregation techniques for wireless sensor networks: a survey," IEEE Wireless Communications, 14(2), pp. 70–87, April 2007.

[3] Dnyaneshwar Mantri, Neeli R Prasad, and Ramjee Prasad, "Two Tier Cluster-based Data Aggregation (TTCDA) in Wireless Sensor Network," Springer Journal of Wireless Personal Communications, 75(4), pp. 2589–2606, Nov. 2013.

[4] Jamal N. Al-Karaki, Raza Ul-Mustafa, and Ahmed E. Kamal, "Data aggregation and routing in Wireless Sensor Networks: Optimal and heuristic algorithms," Computer Networks, 53(7), pp. 945–960, May 2009.

[5] Dnyaneshwar Mantri, Neeli Rashmi Prasad, and Ramjee Prasad, "Mobility and Heterogeneity aware Cluster-Based Data Aggregation for Wireless Sensor Network," Springer Journal of Wireless Personal Communication, 86(2), pp. 975–993, Jan. 2016.

[6] Dnyaneshwar Mantri, Neeli R Prasad, and Ramjee Prasad, "Grouping of Clusters for Efficient Data Aggregation (GCEDA) in Wireless Sensor Network," 3rd IEEE International Advance Computing Conference (IACC-2013), pp. 132–137, Feb. 2013.

[7] Chuan-Yu Cho, Chun-Lung Lin, Yu-Hung Hsiao, Jia-Shung Wang, and Kai-Chao Yang, "Data Aggregation with Spatially Correlated Grouping Technique on Cluster-Based WSNs," 2010 Fourth International Conference onSensor Technologies and Applications (SENSORCOMM), pp. 584–589, July 2010.

[8] Prateek Mathur, Rasmus Hjorth Nielsen, Neeli R Prasad, Ramjee Prasad, "Conasense at nanoscale: Possibilities and challenges," Book, Role of ICT for Multi-Disciplinary Applications, Vol 47, River publishers, 2016.

[9] Wenbo He, Xue Liu, and Klara Nahrstedt, "PDA: Privacy preserving data Aggregation for information Collection," ACM Transactions on Sensor networks, 8(1), Article 6, pp. 1–22, Aug. 2011.

[10] Huifang Chen, Hiroshi Mineno, and Tadanori Mizuno, "Adaptive data aggregation scheme in clustered wireless sensor networks," Computer Communications, 31(15), pp. 3579–3585, Sept. 2008.

[11] Sabrina Sicari, Luigi Alfredo Grieco, Gennaro Boggia, and Alberto Coen-Porisini, "DyDAP: A dynamic data aggregation scheme for privacy aware wireless sensor networks," Journal of Systems and Software, 85(1), pp. 152–166, Jan. 2012.

[12] Ozdemir, S., & Xiao, Y. "Secure data aggregation in wireless sensor networks: A comprehensive overview," Computer Networks, 53(12), 2022–2037. 2009, doi:10.1016/j.comnet.2009.02.023.

[13] L. Zhu, "Secure Data Aggregation in Wireless Sensor Networks" Secure and Privacy-Preserving Data Communication in Internet of Things, Springer Briefs in Signal Processing, 2017. DOI 10.1007/978-981-10-3235-6_2.

[14] Tang, J., Liu, A., Zhao, M., & Wang, T. (2018). "An Aggregate Signature Based Trust Routing for Data Gathering in Sensor Networks". Security and Communication Networks, 2018, 1–30. doi:10.1155/2018/6328504.

[15] Chitnis M, Pagano P, Lipari G, and Yao Liang, "A Survey on Bandwidth Resource Allocation and Scheduling in Wireless Sensor Networks," International Conference on Network-Based Information Systems, 2009. NBIS '09. pp. 121–128, Aug. 2009

[16] Dnyaneshwar S. Mantri, Neeli Rashmi Prasad, and Ramjee Prasad, "Bandwidth Efficient Hybrid Synchronization for Wireless Sensor Network," 4th International Conference on Advances in Computing, Communications and Informatics (ICACCI2014), pp. 2108–2113 Aug. 2014.

[17] Woo, A and Culler, D. A, "Transmission Control Scheme for Media Access in Sensor Networks," In Proceedings of the Seventh International ACM Conference on Mobile Computing and Networking (MOBICOM) pp. 1–15, 2001.

[18] G. Gupta, M. Younis, "Load-Balanced Clustering in Wireless Sensor Networks," proceedings of International Conference on Communications (ICC 2003), vol. 3, pp. 1848–1852, May 2003.

[19] Rout R. R., Ghosh S.K., Chakrabarti S., "Network coding-aware data aggregation for a distributed Wireless Sensor Network," 2009 International Conference on Industrial and Information Systems (ICIIS) pp. 32–36, Dec. 2009.

[20] Brian Honan, BH Consulting," LAYERED SECURITY, PROTECTING YOUR DATA IN TODAY'S THREAT LANDSCAPE", A White Paper.

3

Reliable and Secure LPWAN Communication in Interference Environments

Viktor Stoynov, Dimitriya Mihaylova, Zlatka Valkova-Jarvis, Georgi Iliev and Vladimir Poulkov

Technical University of Sofia, Bulgaria
E-mail: vstoynov@tu-sofia.bg; dam@tu-sofia.bg; zvv@tu-sofia.bg; gli@tu-sofia.bg; vkp@tu-sofia.bg

3.1 Introduction

In the next few years the Internet of Things (IoT) and Wireless Sensor Networks (WSNs) are aiming to change the whole world by wirelessly connecting various categories of physical objects. These types of networking will give rise to an enormous growth in new wireless technologies and standards, and will also generate huge amounts of traffic due to the usage of sensors. For this reason, the establishment of a large variety of Low Power Wide Area Networks (LPWANs) is of the utmost importance. These networks are based on the deployment of communications between devices in order to perform monitoring and control tasks across extensive geographical areas with minimal human intervention [1]. In this context, the realization of LPWAN applications faces several challenges. Among these are the use of a limited spectrum, coexistence with other technologies operating in the same band, mobility, scalability, coverage, and security [2]. In addition, data rates and real-time communication need to be improved.

The major problems of LPWANs are interference and the significant attenuation of signals in the industrial, scientific and medical (ISM) range, primarily due to the long distance between sensors and gateways. Together with the interference produced by heterogeneous and homogeneous devices, the coexistence of different networks may also be a source of interference

Security within CONASENSE Paragon, 35–58.

for the technologies operating in the ISM bands. This interference, coupled with the many other problems related to ISM's dynamic nature, can seriously impact quality of service (QoS). Various studies have already examined the influence of interference on system performance. However, it is still necessary to study the particularities of transmitting signals, channel features and parameters in detail. To a large extent, the presence of interference largely depends on the coexistence of many different IoT technologies. For example, one real challenge is the analysis of different LPWAN implementation scenarios operating in one frequency range with non-orthogonal carrier frequencies. The interference described above usually occurs accidentally and is defined as *unintentional* (or *accidental*) interference.

Since confidential information is usually exchanged over LPWANs, one emerging issue is the enhancement of their security. Although end-to-end cryptography is supported by most of the available LPWAN technologies, it is associated with increased complexity, latency and energy consumption. Moreover, cryptography techniques are not sufficient to overcome attacks initiated over lower layers of the model architecture, such as traffic analysis, collisions, resource exhaustion or unfairness resource allocation. Another attack whose mitigation is not considered in the LPWANs is denial-of-service (DoS) at the physical layer, also known as a jamming attack.

The jamming attack as a means of disrupting the security of an LPWAN system involves the deliberate transmission of interfering signal aiming, on the one hand, to reduce the signal-to-interference-plus-noise-ratio (SINR) at the receiver end and, on the other hand, to occupy the channel and prevent legitimate access. As jamming is a deliberate malicious intervention, it is also referred to as *intentional interference*.

In this chapter, the influence of unintentional and intentional interference on the system's performance of LPWANs is discussed. The major applications and unresolved challenges concerning the LPWANs are presented and data analytics-based approach is presented in order to achieve an efficient unintentional interference management. Moreover, a brief summary of the methods, used on the physical layer against jamming attacks, is presented together with a comparison between them based on several of their major properties.

3.2 LPWAN Applications and Unresolved Challenges

The ever-increasing attractiveness of LPWANs in industrial and research communities is mainly due to their low energy consumption, low-cost communication characteristics and long-range communication capabilities. Generally, these networks can provide coverage within 10–40 km in rural zones and 1–5 km in urban zones [3]. Additionally, the LPWANs are extremely energy efficient, with up to 10 or more years of battery lifetime, and low-cost, at around the cost of a single radio chipset [4]. These characteristics and capabilities of LPWANs have inspired engineers to realize numerous experimental studies on their performance in outdoor and indoor environments. The up-to-date LPWAN technologies usually use gateways, referred to as concentrators or base stations (BSs), to serve end-devices. In this context the end-devices communicate directly with one or more gateways. This is the major difference between traditional WSNs and LPWANs. This type of topology significantly simplifies the coverage of large areas, even spanning an entire nation, by taking advantage of already deployed cellular network infrastructure.

Many LPWA technologies such as SigFox, LoRa, and Ingenu RPMA (based on Random Phase Multiple Access–RPMA), are proprietary-based networks and operate in unlicensed ISM frequency bands. Accordingly, the proximity of other networks using the same bandwidth may cause high levels of interference. With the aim of network performance enhancement and the realization of long-range communication, several LPWAN technologies use the sub-1 GHz bands, which are highly fragmented. Contrary to the concepts of frequency allocation adopted in SigFox and LoRa networks, Ingenu RPMA can support a large coverage network because of its receiver sensitivity of -145 dBm which is acceptable worldwide without being restricted, based on the policy regulations for the 2.4 GHz band. In Table 3.1 a comparison of different LPWA solutions is shown. The choice of technologies to be presented is based on the deep analysis of publicly available data on each technology, along with the amount of published papers focused on each of the applications that each technology targets. Obviously, most LPWAN technologies use the sub-1 GHz ISM band and each of them can be used in different application scenarios.

LPWAN applications are intended to be implemented in almost all Business and Industrial sectors, together with other areas such as Communal Living, Service, Science and Education, etc. The principal application of LPWANs is in the realization of Smart City. Two important examples of such an application are Waste Management and Smart Lighting. The latter aims to substantially lower street-lighting costs by varying the intensity of the lighting in line with the needs of the environment. It also manages to reduce maintenance costs through intelligent fault monitoring.

Table 3.1 Comparison of different LPWAN technologies

	LoRa	Sigfox	Weightless			Dash-7	Ingenu RPMA	Telensa
			–W	–N	–P			
Band	Sub-1 GHz ISM	Sub-1 GHz ISM	TV White Spaces	Sub-1 GHz ISM	Sub-1 GHz ISM	Sub-1 GHz ISM	2.4 GHz ISM	Sub-1 GHz ISM
Channel width	EU: 8×125 KHz; US: 64×125 KHz/ 8×125 KHz	UNB–100 Hz – 1.2 KHz	5 MHz	UNB–200 Hz	12.5 KHz	25 kHz or 200 kHz	1 MHz (40 channels)	?
Range	2–5 km (urban); 15 km (suburban); 45 km (rural)	Up to 10 km (urban); 50 km (rural)	5 km (urban)	3 km (urban)	2 km (urban)	0–5 km	>500 km (LoS)	1 km (urban)
Modulation	CSS; US: LoRa; EU: LoRa and GFSK	UNB DBPSK (UL), GFSK (DL)	16-QAM, BPSK, QPSK	UNB DBPSK	GMSK, QPSK	GFSK	RPMA-DSSS (UL), CDMA(DL)	UNB FSK
Data rate	0.3–37.5 kbps (LoRa), 50 kbps (FSK)	100 bps (UL), 600 bps (DL)	1 kbps–10 Mbps (UL and DL)	30–100 kbps (DL), no UL	200 bps–100 kbps (UL and DL)	9.6, 55.6, 166.7 kbps	78 kbps (UL), 19.5 kbps (DL)	62.5 bps (UL), 500 bps (DL)

Transmition power	EU: < +14 dBm; US: < +27 dBm	10 µW to 100 mW	17 dBm	17 dBm	17 dBm	433 MHz: +10dBm; 868/915 MHz: +27dBm	Up to 20 dBm	?
Bi-directional transmission	Yes, depends on mode	No	Yes	Uplink only	Yes	Yes	Yes	Yes
Topology	Star on Stars	Star	Star	Star	Star	Node-to-node, Tree, Star	Star, tree	Star
Software Updates	Yes, Over the Air	No	?	No	Yes, Over the Air	?	Yes, Over the Air	Yes, Over the Air
Authentication and encryption	AES 128b	No support	AES 128b	AES 128b	AES 128/256b	AES 128b	16B hash, AES 256b	No support
Governing body	Lora Alliance	Sigfox	Weightless SIG			Dash Alliance	Ingenu	Telensa

A further example of a specific application is Connected Vehicles. The newest vehicles have networking capability and come equipped with processors and sensors in order to provide an enhanced driving experience through such factors as improved road sharing, accident reporting, parking detection etc. Due to their long-range communication ability, low power, low cost, and mobility, LPWANs can be used for Transportation and Logistics applications.

Another major market for LPWAN applications is health monitoring. Currently, the Healthcare sector has seen the extensive adoption of short range wireless technologies such as ZigBee, WiFi, 6LowPAN, together with cellular technologies – LTE, for instance. Nevertheless, the increased number of sensors and the presence of interference will greatly impede the performance of these networks. In this context LPWANs become more and more an alternative communication solution for Healthcare applications due to the high cost of cellular technologies and the limitations of short-range wireless ones.

LPWANs are currently in the intensive development phase and many problems may be defined. Some of them already have solutions but these are not efficient enough, while the solutions of others are still under development, and other problems which have not yet been addressed.

In addition, there are several so far unresolved challenges of LPWANs, in brief:

Inter-technology Communication

The rapid evolution of LPWAN technologies has contributed to the increase in the number of coexisting LPWANs in the same geographical areas. In such circumstances interference management and inter-network coordination become particularly important issues. A major challenge can arise when LPWA technologies created by different vendors must communicate with each other. Recently, cross-technology-communication (CTC) without additional hardware assistance has been studied for communication across WiFi, ZigBee, and Bluetooth devices. However, future research is necessary to enable CTC in LPWANs [5].

High Data Rate Support

The typical guaranteed data rate of LPWAN technologies ranges from 1 kbps to 100 kbps. The widely known narrowband communication used in several

LPWAN technologies offers long transmission range at the price of low data rates. The advent of aerial imagery systems, which comprise drones and cameras for richer sensor data from farms, necessitates the use of increased bandwidth in agricultural IoT [6]. Sooner or later, numerous IoT applications will change in order to include several use cases, such as video streaming, requiring very high data rates. Different methods and algorithms for high data rate support in LPWANs need to be considered. Future re-search directions to enable high data rates comprise: support of different modulation techniques, implementing different approaches used in technologies like WiFi, and designing new hardware to support multiple physical layers offering different data rates.

Inter-LPWAN Coordination and Sharing of Available Resources

Resource sharing and inter-LPWAN coordination are among the ways to achieve scalability improvements. LPWANs all operate in unlicensed bands, giving anyone the opportunity to activate their own LoRaWAN, 802.15.4 or HaLow network, for example. In the case of any given type of LPWAN technology, this can result in a set of several co-located networks without any coordination, leading to a reduction in scalability, due to interference. In order to share resources, multiple networks need to cooperate, requiring cross-coordination to reduce interference between co-located single-technology LPWANs. Up to now, virtualization of wireless networks has been focused on 3GPP LTE and IEEE 802.11 [7]. Additional research is desirable to design virtualization solutions and management methods for LPWAN technology, and to investigate the improvement in scalability that can be attained by applying suitable coordination mechanisms.

Technology Co-existence and Interference Management

The extreme attractiveness of LPWAN applications has triggered a new challenge – coexistence. Due to the deployment of many separate networks in close proximity, mutual interference has to be controlled in order to maintain satisfactory levels of operational status. This challenge will lead to spectrum overcrowding, something which LPWANs are currently not geared up for. Current studies of LoRa and SigFox demonstrate that coexistence leads to severe performance degradation. Due to interference, the overall throughput does not change significantly when several LoRa networks coexist. Current coexistence management for WiFi and Bluetooth cannot be implemented well in LPWANs. This is mainly due to the fact that LPWAN devices, owing to their large coverage domains, can be subject to an unparalleled number of hidden terminals.

One opportunity for research is the recognition and identification of other wireless technologies by the use of spectrum information. This can be achieved through intelligent spectrum monitoring and the use of an efficient spectrum sensing method or dedicated hardware, combined with machine learning techniques to identify those technologies which may be interfering [8]. These techniques are discussed in Section 1.3.

When the presence of different networks is considered, the unintentional interference may be categorized as Inter-network Interference (IrnI) or Intra-network Interference (IanI). The transmissions in the first category are generated by end-devices (sensors) belonging to two or more distinct LPWANs. IrnI can be mitigated by flexible approaches such as spectrum sensing, radio environment maps or a spectrum occupancy database. When the interference is generated between sensors belonging to the same LPWAN it is termed IanI, or self-interference. Mitigation of this sort of interference can be achieved through collision-recovery and collision-avoidance schemes. Two types of IrnI are known to exist: homogeneous (HoI) and heterogeneous (HeI). The first type occurs when two or more networks are utilizing the same radio technology and the second type when, for example, different modulation schemes are used [9]. Several recently proposed approaches to interference identification based on spectrum utilization monitoring used to decrease the negative impact of technology co-existence in LPWANs are presented and discussed in the next section. In the context of LoRaWAN – the leading LPWA technology, there are two additional types of interference that may occur due to the usage of CSS. They are Co-SF Interference (CSFI), where end-devices with the same spreading factors are subject to collisions, and Inter-SF Interference (ISFI), where end-devices with different SFs experience collisions. The results in [10] show the non-negligible impact of SFs' imperfect orthogonality, along with the negative impact of SF allocations on the overall throughput.

Depending on its occurrence and bandwidth occupancy, intentional interference can be divided into four major types [11, 12] – wideband, single-tone, multi-tone, and single-tone jamming with frequency hopping. While wideband jamming covers the whole spectrum, single-tone jamming consists in sending a narrowband signal with a specific central frequency. In single-tone jamming with frequency hopping, the central frequency of the narrowband interfering signal varies quickly in time in order to affect more sections of the spectrum. Another way to invade larger frequency bandwidth is with intentional interference, called multi-tone, where the power of the jammer is distributed between several narrowband signals with differing central frequencies, forming parts of the spectrum that are jammed simultaneously. Although jamming the entire spectrum may have a stronger impact on the system's performance, the resource utilization needed to initiate such an attack is usually too high to be practical, making single-tone jamming the type of intentional interference preferred by jammers.

Several classifications of single-tone jamming with respect to its behavior and duration exist in the literature [11–13], and can be briefly summarized as follow:

- Constant – permanent transmission of interference with random signals;
- Deceptive – permanent transmission of interfering signals which resemble regular frames;
- Random – interference transmission at random time intervals;
- Reactive – jamming initiated only when legitimate transmission is detected;
- Adaptive – an idealized jamming scenario, where the power of the interfering signal varies depending on the estimated legitimate channel gain;
- Intelligent – jamming the upper layers of the protocol stack, based on knowledge of the protocols used there.

In LPWANs, DoS at the physical layer can be initiated during the data transmission phase; the channel estimation phase, where it is known as pilot contamination attack [14]; or during the LPWAN join procedure of the end nodes, to disable their connection to the network [15]. Numerous mechanisms for jamming mitigation that work on the different layers of the reference model are available in the literature [16–18]. However, for the LPWAN applications, where very small devices with hardware and resource constraints are incorporated, methods from the upper layers can be unsuitable due to the additional overhead, computational complexity and energy consumption. For that reason, lightweight and simple techniques from the Physical Layer Security (PLS) are most appropriate to be used for jamming attack suppression in LPWANs [19–22]. In Section 1.4 of this chapter an overview of different methods for jamming mitigation at the physical layer which can be implemented in LPWANs is given.

3.3 Unintentional Interference Management in LPWANs

As already shown in Section 1.2, one of the key limitations in wireless radio communications remains the scarcity of available spectrum. This often has a negative impact on mobile connectivity and can cause network congestion and mutual interference between networks using overlapping frequency bands. This problem is often seen in unlicensed ISM bands. Several LPWAN technologies, like LoRaWAN and Sigfox, have moved away from the over-crowded 2.4 GHz band in order to use the unlicensed spectrum in the sub-1GHz frequency range. The first reason for this shift is increased range, which allows the realization of a complete new set of low-energy IoT use cases. Furthermore, the level of utilization of the unlicensed sub-1GHz spectrum bands is very low and their use may help to avoid the interference issues of the 2.4 GHz band. The main disadvantage with the operation and management of unlicensed networks is the completely uncoordinated deployment, which may cause significant performance degradation. With the intention of mitigating the negative impact, advanced and intelligent management, coordination and collaboration mechanisms that mitigate interference and improve coexistence among different networks (using the same and different technologies) operating in the unlicensed sub-1GHz frequency ranges need to be put forward and investigated in depth.

Many existing publications propose different approaches for technology coexistence in the very widely-used 2.4 GHz band [23]. Conversely, the effects of wireless network coexistence in the sub-1GHz band are still poorly studied. However, similar properties are to be expected as a result of the large number of LPWAN technologies operating in the same unlicensed band, especially since their low-power nature makes these networks very susceptible to interference from outside sources. Thus, it is essential to implement techniques which can rapidly identify and properly categorize the interfering technologies. Moreover, improvements in the usage of information about interference originating from different locations and selection of suitable mitigation technique need to be considered.

Many of the already existing approaches for resource scheduling and interference mitigation are not applicable to sub-1 GHz LPWANs [24], and the improvement of hardware intelligence is not possible, due to the usage of very low-cost devices (sensors). On the other hand, the frequently-used frequency hopping technique is not appropriate, due to the scarcity of available spectrum. Due to LPWANs' technology-specific nature, high resource and energy requirements, the use of spectrally efficient modulation schemes or 802.11ax-based approaches for interference control are unsuitable [25].

The first possible course for sub-1GHz LPWAN optimization is the investigation of the use of adaptive power control as a scalability feature in LPWAN networks. In fact, the range of transmissions can be controlled based on the distance between sender and receiver, by carefully choosing the accurate power level. In this way reduction of interference between stations can be achieved. This is a very important feature, especially in densely deployed areas where a short transmission range is sufficient to reach the next device (sensor). On the other hand, the interference between co-located single-technology LPWANs may be reduced by cross-coordination and cooperation between multiple networks. This technique primarily involves intelligent resource sharing and the coordination of scheduling between LPWANs. The existing studies investigating the problem of wireless networks virtualization and cooperation have mainly focused on 3GPP LTE and IEEE 802.11 [26–28]. Here, the creation of intelligent virtualization solutions and management techniques intended to be used in LPWAN networks is needed in order to achieve high levels of scalability and immunity from interference.

In terms of system reconfiguration, so as to enhance network adaptability, certain approaches may be used, such as automatic selection of more robust coding schemes, limiting packet sizes, increasing error correction codes, or simply defining an ad-hoc and very granular cross-technology TDMA (Time Domain Multiple Access) scheme.

Another suitable approach for interference identification and mitigation in LPWANs is spectrum sensing and the use of dedicated spectrum sensing devices. In order to accurately recognize the available spectrum and the types of LPWAN technologies that are currently present in an area, based on spectrum and network performance information like packet loss rates, typical sizes of error bursts, etc., machine learning approaches must be developed.

One of the main emerging paradigms for network performance improvement is RF Data Analytics. This is a promising option for the optimization of wireless networks such as LPWANs, due to its capability of improving spectrum utilization along with quality of experience (QoE) in a user-centric manner. From a theoretical point of view, RF data is considered to be either time-domain baseband in-phase and quadrature (IQ), or frequency-domain (spectrum). RF data is created using radio receivers that are able to cover a wide band but which operate in one narrow band at a time.

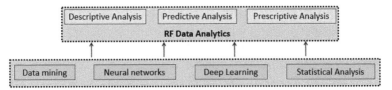

Figure 3.1 Basic RF DA architecture.

Three different kinds of DA have already been defined in the literature, as depicted in Figure 3.1. The first, Descriptive Analysis (DesA), comprises Data Modelling, Visualization and Regression. These three stages aim to prepare the data for subsequent analysis. After the data is collected, a representation of this data in a meaningful form is carried out in order to detect a simple trend. One example of a descriptive RF data analytics service and application is spectrum monitoring for operation estimation. In this case, the purpose of the monitoring is to start with information about the systems known to be operational and provide feedback for spectrum management, i.e., close the loop, by confirming that the devices are operating as authorized or by providing evidence that they are not [29].

As soon as the DesA is completed, a Predictive Analysis (PA) has to be done. This involves Data Mining (DM) and Predictive Modelling (PM). The first step, DM, aims to extract different patterns from the whole of the data already collected. The major aim of the second step, PM, consists in intelligent recognition of the trends and realization of different prediction techniques. When the PA is successfully completed it is time for the last step of DA – the Prescriptive Analysis (PrA). On the basis of the analysis carried out so far, the main purpose of PrA is decision making and the optimization of the entire process [30].

Spectrum utilization forecasting and interference identification are two promising applications of RF data predictive and prescriptive analytics. The practical implementation of PA is based on ML tools, with the purpose of developing models on the basis of past data for future prediction. The study in [31] discusses a survey of the emerging ML approaches as essential components of DA, together with some of the challenges in this scientific area. A practical implementation of spectrum utilization forecasting based on ML algorithms on RF data was presented in [32]. The data was obtained after long-term spectrum monitoring at an air traffic control station in Bulgaria. It becomes clear that the application of proper ML-based data analytics to the data obtained from long-term spectrum monitoring results in satisfactory forecasting mainly for channels which are active more than 50% of the time.

In order to implement intelligent RF DA in the context of LPWANs, several basic requirements must be satisfied:

- The collection of RF data must be based on significant long-term monitoring;
- The process of spectrum data collection must be characterized by high flexibility and intelligence;
- Distinct spectrum holes can be found by implementing various dynamic resource allocation strategies;
- The RF DA must have the ability to predict, in an intelligent manner, the variations pertaining to the spectrum utilization;
- A scalable platform must be implemented in order to implement intelligent methods and algorithms for analysis and learning for the data collected from long-term monitoring;
- The platform used for RF DA must be characterized by a high level of control and the ability to intelligently reconfigure radio parameters;
- In order to carry out high quality interference monitoring and prediction, statistical techniques and ML algorithms must be proposed;
- DA must be used also in order to implement new improved approaches for intelligent and efficient spectrum utilization.

3.4 Physical Layer Security Methods for Jamming Suppression

In this section, different methods for jamming mitigation used at the physical layer are presented. These methods are divided into three major groups: spread spectrum techniques, non-adaptive, and adaptive jamming filtering at the gateway. In order to illustrate their application in concrete practical examples, the methods are discussed when employed in specific networks and systems, however they can be also applied in LPWANs.

3.4.1 Spread Spectrum Techniques

A commonly-used method to combat the influence of noise, intentional and unintentional interference in wireless systems is spread spectrum modulation. In essence, spreading techniques operate so that the narrowband communication signal is expanded into a broadband frequency spectrum through specific processing at the end node. For de-spreading purposes, similar processing is conducted at the gateway in order to demodulate the signal into its original form.

Two main challenges are common to all spread spectrum techniques. To achieve proper demodulation of the received signal, synchronization is needed between the end node and the gateway. Another important issue, which may hamper the exploitation of these types of modulation, relates to spectrum occupancy. Spreading the narrowband signal into the entire bandwidth results in inefficient spectrum utilization.

Various spread spectrum techniques exist and in this chapter the Direct Sequence Spread Spectrum (DSSS) and Frequency Hopping Spread Spectrum (FHSS) are discussed, as they are the most frequently used, and the Chirp Spread Spectrum (CSS), used by LoRaWAN, is examined as a novel physical layer technology.

Frequency Hopping Spread Spectrum

FHSS is a spreading modulation where the carrier frequency alters quickly in time so that interference at a certain frequency has negligible effect on the legitimate transmission [33]. The frequency variations are accomplished according to a pseudo-random pattern, known at both gateway and end node, which can be predetermined through PLS methods for key distribution. Unless the attacker discovers the hopping sequence, interference at an individual frequency will affect legitimate signals only in the short time interval where both the frequencies coincide.

Direct Sequence Spread Spectrum

Spreading with DSSS [34] is carried out through the use of a random sequence of pseudo-noise chips expressed by rectangular pulses of values 1 and -1. In this way, the original signal is expanded into the entire bandwidth and assumes the spectral characteristics of a wideband noise signal, making legitimate transmissions undetectable by a malicious node. Since the power spectrum density of the signal is spread over a large bandwidth, narrowband jamming has an influence on only a relatively small part of it. As with FHSS, the pseudo-noise sequence must be arranged between the end node and gateway in advance.

Chirp Spread Spectrum

In CSS the signal is modulated through chirp pulses, where every chirp represents a sinusoid with linearly- or exponentially-changing frequency.

When the frequency of the chirp increases in time, it is called up-chirp, and down-chirp when decreasing. Spreading in a wider frequency band is achieved by multiplication of the data with a base chirp [35]. In CSS, synchronization is needed to deal with the time offset that occurs during the transmission. The abilities of CSS to overcome multi-path fading and the Doppler Effect make it suitable for long range communications. It is included as a physical layer of the LoRaWAN technology.

3.4.2 Non-adaptive Jamming Filtering

Regularized Zero-Forcing Filtering

A filter at the gateway may be constructed according to the algorithm proposed in [36] for regularized zero-forcing (RZF) filtering. For this purpose, the following assumptions must be made: first, a pilot sequence, which is orthogonal to those used by the end node will remain unused, and second, the power and large-scale fading of the jamming signal are already known.

During the pilot session the legitimate channel is estimated. The channel gain of the jammer is then obtained by zero-forcing projection of the received training sequence over the unused pilot sequence. When the estimations of both the legitimate and jamming channels are calculated – $\hat{h}$ and $\hat{g}$, correspondingly, the RZF filter, a_{RZF}, is designed according to (3.1), where η is the regularization factor, M is the number of antennae at the gateway and $\mathbf{I}_M$ is an identity matrix of size M:

$$a_{\mathrm{RZF}} = \left(\hat{g}\hat{g}^H + \eta\mathbf{I}_M\right)^{-1}\hat{h}. \tag{3.1}$$

The authors in [29] investigate two types of the RZF filter depending on the regularization factor in use – MMSE-type RZF filter, based on Minimum Mean Square Error (MMSE) estimation, and ZF-type RZF filter, based on zero-forcing (ZF) for jamming elimination.

Blind Jamming Mitigation Filtering

The Blind Jamming Mitigation (BJM) algorithm, introduced in [37], establishes a linear spatial filter $\mathbf{P}$ only by the knowledge of the pilot sequences of L number of pilots, sent by the end node, $\left[\tilde{X}(1), \tilde{X}(2), ..., \tilde{X}(L)\right]$, and the one, received at the gateway – $\left[\tilde{\mathbf{Y}}(1), \tilde{\mathbf{Y}}(2), ..., \tilde{\mathbf{Y}}(L)\right]$. This is achieved by minimizing the Mean Squared Error (MSE) of the estimated training signals, following (3.2):

$$\mathbf{P} = \left[\sum_{l=1}^{L} \tilde{\mathbf{Y}}(l) \, \tilde{\mathbf{Y}}(l)^{H} \right]^{\dagger} \left[\sum_{l=1}^{L} \tilde{\mathbf{Y}}(l) \, \tilde{X}(l)^{H} \right]. \qquad (3.2)$$

where $(.)^{\dagger}$ denotes a pseudo-inverse matrix.

IEEE 802.15.4 Digital Filtering

Although DSSS is applied in the IEEE 802.15.4 standard, the incorporated 2-byte Cyclic Redundancy Check (CRC) procedure represents a weakness that can be exploited to initiate a DoS attack at the physical layer. In a case where intentional interference is directed against the CRC procedure, jamming of one symbol per packet will result in different computations of the CRC code at the end node and gateway, making the transmission unsuccessful. To counter such an intrusion, the authors of [38] recommend the use of a simple Finite Impulse Response (FIR) high-pass filter of low order. The proposed filter removes jamming whose central frequency is the same as that of the legitimate transmission, at the baseband. However, if both the frequencies differ, which is the real-world case, the method is not effective. Furthermore, in non-attack scenarios, this filtering technique significantly reduces the quality of the signal received at the gateway. To surmount these shortcomings, the authors in [39] propose the use of a filter bank, where adaptive filter selection is provided – this is described in the next subsection.

3.4.3 Adaptive Jamming Filtering

IEEE 802.15.4 Adaptive Selection of Digital Filtering

The solution proposed in [39] is based on the IEEE 802.15.4 digital filter, described in the previous subsection, and represents a combination of several such filters, a packet error detection block, and a controller. The Packet Delivery Ratio (PDR) of the signal received at the gateway is calculated at the packet error detection block and, depending on the value obtained, the controller decides if filtering is needed and which filter from the bank best suits the existing interference. Thus, the selection of the appropriate filter from the bank can be adapted to the central frequency of the jamming signal and, since filtering is not carried out in the absence of an attack, the signal quality remains unaffected. In spite of the benefits of the IEEE 802.15.4 adaptive filter's selection algorithm, its application in LPWANs may be precluded by the increased complexity introduced by the controller and the PDR computations conducted at the MAC layer.

GPS Complex Notch Filtering

A module composed of three complex adaptive notch filters (CANF) is suggested in [40] for GPS applications but can be also be adopted by LPWANs. The filters are of first, second and third order and, depending on the value of the Jamming-to-Noise Ratio (JNR), one of them is selected for jamming filtering. In the case of a low JNR, a first order filter is used, while higher interference power is suppressed by a higher order filter from the CANF module. The JNR is indirectly determined by the variance of the complex adaptive coefficient b_0, which relates to the central frequency of the interference. The first order filter from the CANF module is described by the following transfer function, where k_α assigns the width of the notch:

$$H_{CNF}(z) = \prod_{K=1}^{K} \frac{1 - b_0 z^{-1}}{1 - k_\alpha b_0 z^{-1}}. \tag{3.3}$$

The adaptive filter equation, by which the central frequency of the notch is adapted to the one of the jamming signal, is given in (3.4):

$$b_0[n] = b_0[n-1] + \mu \times 4x_0[n](x_e^*[n-1]), \tag{3.4}$$

where the moving average and autoregressive outputs of the filter are $x_0[n]$ and $x_e[n]$, respectively, μ is the step size and $(.)^*$ stands for complex-conjugate.

Despite being an effective solution, both the CANF module and the algorithm to choose the most appropriate filter increase the system's complexity and may result in an inefficient LPWAN application.

Adaptive Complex Narrowband Filtering

In [41] a complex Low Sensitivity (LS1) filter section of the first order is initially described to be used for an adaptive filter design. Its transfer functions are of band-pass type and are given in (3.5):

$$H_{RR}(z) = H_{II}(z) = \beta \frac{1+2\beta \cos\theta z^{-1}+(2\beta-1)z^{-2}}{1+2(2\beta-1)\cos\theta z^{-1}+(2\beta-1)^2 z^{-2}};$$

$$H_{RI}(z) = -H_{IR}(z) = \beta \frac{2(1-\beta)\sin\theta z^{-1}}{1+2(2\beta-1)\cos\theta z^{-1}+(2\beta-1)^2 z^{-2}}.$$

(3.5)

In [42] the adaptive complex notch filter is proposed to be used for jamming attack mitigation in resource-constrained wireless networks. Both the bandwidth and central frequency of the filter can be adaptively changed according to the characteristics of the intentional interference by tuning the coefficients β and θ, correspondingly. In order to adjust the central frequency of the notch to conform with that of the jamming, the Least Mean Square (LMS) algorithm is used to form the adaptive equation:

$$\theta(n+1) = \theta(n) + \mu Re[e(n)y^{'*}(n)],$$

(3.6)

where μ is the step size, $e(n)$ is the output of the notch, $y'(n)$ is the derivative of the band-pass output with respect to θ.

The adaptive complex narrowband filter is suitable for LPWANs due to its adaptivity, low order providing small power consumption, and low hardware and computational complexity.

3.4.4 Comparison of the Physical Layer Methods for Jamming Suppression

A comparison of the main properties of the methods discussed for jamming mitigation is given in Table 3.2. As can be seen from the table, filtering techniques surpass the spread spectrum approaches in bandwidth efficiency, lack of synchronization, and lack of processing at the end node. However, some of the filters described are of very high complexity, work with the estimated Channel State Information (CSI) or have special requirements concerning the pilot transmission phase, which could be considered as their main drawbacks. According to the table, the adaptive complex narrowband filter of LS1 type has the simplest implementation among the methods discussed, which makes it the most appropriate choice for LPWAN applications.

Table 3.2 Comparison of the techniques used at the physical layer for jamming suppression

No	Features Compared	Jamming Mitigation Technique						
		Spread Spectrum	RZF	BJM	Digital Filter for IEEE 802.15.4	IEEE 802.15.4 Adaptive Selection	GPS Notch Filtration	LS1 Adaptive Filtration
1	Additional processing at the receiver	✓	✓	✓	✓	✓	✓	✓
2	Additional processing at the transmitter	✓	X	X	X	X	X	X
3	Synchronization needed	✓	X	X	X	X	X	X
4	Increased bandwidth needed	✓	X	X	X	X	X	X
5	Performance loss in non-attack scenarios	X	X	X	✓	X	X	X
6	High computational complexity	✓	✓	X	X	✓	✓	X
7	CSI needed	X	✓	X	X	X	X	X
8	Large number of pilots needed	X	✓	✓	X	X	X	X
9	Purposely unused pilot sequence needed	X	✓	X	X	X	X	X

3.5 Conclusion

LPWANs provide important improvements in terms of power consumption, coverage, deployment cost and pricing over cellular and other connectivity technologies. Consequently, the widespread adoption of unlicensed LPWAN solutions is anticipated in the coming years. The analysis in this chapter clearly shows the advantages of sub-1 GHz LPWAN implementation, as well as the unresolved challenges, especially when considering interference environments. In this context, cooperation between networks and the implementation of RF Data Analytics seem to be two possible approaches when solutions for growing spectrum scarcity and technology coexistence-based interference need to be found. Future work concerning these problems may focus on the realization of distinct RF DA based algorithms and methods. Additionally, research can be considered into the actual spectrum utilization in the sub-1 GHz and 2.4 GHz ISM band across different scenarios, as well as exploring approaches to the possible realization and application of mechanisms to provide more effective access to these resources.

Since jamming attacks represent a major challenge for the security enhancement of LPWANs, the main physical layer security techniques to counteract them are summarized in this chapter. The analysis of the methods shows that both the classes of the approaches, namely spread spectrum techniques and jamming filtering at the gateway, possess their own advantages and drawbacks. Therefore, in most cases a trade-off must be made between the jamming resistance of LPWAN and the complexity associated with inefficient resource utilization. However, of the methods discussed, the LS1-based adaptive complex narrowband filter appears to be a reasonable solution. In a future work, the performance of the adaptive notch filter with complex coefficients can be investigated. The signal-to-jamming-ratio gain can be analysed when changing the step size and the bandwidth of the filter, in order to study the effect of these parameters on the system's performance.

Acknowledgement

This work was supported by Research Project D-054-2018 funded by the R&D&I Consortium of Sofia Tech Park, Bulgaria.

References

[1] A. Mukherjee. Physical-layer security in the Internet of Things: Sensing and communication confidentiality under resource constraints. *In Proc. IEEE*, 103(10): 1747–1761, 2015.

[2] B. Reynders, W. Meert, and S. Pollin. Range and coexistence analysis of long range unlicensed communication. In *Proceedings of 2016 23rd International Conference on Telecommunications (ICT)*, 1–6, 2016.

[3] M. Centenaro, L. Vangelista, A. Zanella, and M. Zorzi. Long-range communications in unlicensed bands: The rising stars in the IoT and smart city scenarios. *IEEE J. Wirel. Comm.*, 23(5):60–67, 2016.

[4] D. Patel, and M. Won. Experimental study on low power wide area networks for mobile internet of things. In *Proc. of VTC*, 1–5, 2017.

[5] Song Min Kim, and Tian He. Freebee: Cross-technology communication via free side-channel. In *MobiCom, ACM*, 2015.

[6] D. Vasisht and Z. Kapetanovic. FarmBeats: An IoT Platform for Data-Driven Agriculture. In *Proceedings of 14th USENIX Symposium on Networked Systems Design and Implementation (NSDI 17)*, 515–529, 2017.

[7] J. Petajajarvi, K. Mikhaylov, A. Roivainen, T. Hanninen, and M. Pettissalo. On the coverage of LPWANs: Range evaluation and channel attenuation model for LoRa technology. In *Proceedings of the 14th International Conference on ITS Telecommunications (ITST)*, 55–59, 2015.

[8] O. Georgiou and U. Raza. Low Power Wide Area Network Analysis: Can LoRa Scale? In *IEEE Wireless Communications Letters*, 99:1–1 (2017). ISSN: 2162-2337. DOI: 10.1109/LWC.2016.2647247.

[9] L. Stabellini. Design of Reliable Communication Solutions for Wireless Sensor Networks. Licentiate Thesis in Radio Communication Systems Stockholm, Sweden 2009.

[10] V. R. Stoynov, V. K. Poulkov and Z. V. Valkova-Jarvis. Interference Management in LoRaWANs – Overview and Simulation Study. In *Proceedings of 2018 Advances in Wireless and Optical Communications (RTUWO)*, Riga, Latvia, 251–256, 2018

[11] Y. M. Amin and A. T. Abdel-Hamid. Classification and analysis of IEEE 802.15.4 PHY layer attacks. In *Proceedings of 2016 International Conference on Selected Topics in Mobile & Wireless Networking (MoWNeT)*, Cairo, 1–8, 2016.

[12] A. Mpitziopoulos, D. Gavalas, C. Konstantopoulos and G. Pantziou. A survey on jamming attacks and countermeasures in WSNs. In *IEEE Communications Surveys & Tutorials*, 11(4):42–56, 2009.

[13] Y. Zou, J. Zhu, X. Wang and L. Hanzo. A Survey on Wireless Security: Technical Challenges, Recent Advances, and Future Trends. In *Proceedings of the IEEE*, 104(9):1727–1765, 2016.

[14] D. Mihaylova, Z. Valkova-Jarvis and G. Iliev. Detection capabilities of a shifted constellation-based method against pilot contamination attacks. In *Proceedings of 2017 Advances in Wireless and Optical Communications (RTUWO)*, Riga, 56–60, 2017.

[15] S. M. Danish, A. Nasir, H. K. Qureshi, A. B. Ashfaq, S. Mumtaz and J. Rodriguez. Network Intrusion Detection System for Jamming Attack in LoRaWAN Join Procedure, In *Proceedings of 2018 IEEE International Conference on Communications (ICC)*, Kansas City, MO, 1–6, 2018.

[16] R. Muraleedharan, and L. Osadciw. Jamming Attack Detection and Countermeasures In Wireless Sensor Network Using Ant System. In *2006 SPIE Symposium on Defense and Security*, April, 2006.

[17] A. D. Wood, J. A. Stankovic and G. Zhou. DEEJAM: Defeating Energy-Efficient Jamming in IEEE 802.15.4-based Wireless Networks, In *Proceedings of 2007 4th Annual IEEE Communications Society Conference on Sensor, Mesh and Ad Hoc Communications and Networks*, San Diego, 60–69, 2007.

[18] S. Vadlamani, B. Eksioglu, H. Medal and A. Nandi. Jamming attacks on wireless networks: A taxonomic survey, In *Proceedings of International Journal of Production Economics*, 172:76–94, 2016.

[19] Y. Wu, A. Khisti, C. Xiao, G. Caire, K. Wong, and X. Gao. A Survey of Physical Layer Security Techniques for 5G Wireless Networks and Challenges Ahead. In IEEE Journal on Selected Areas in Communications, vol. 36, no. 4, pp. 679–695, April 2018.

[20] Y. Zou, J. Zhu, X. Wang, and V. C. M. Leung. Improving Physical-Layer Security in Wireless Communications Using Diversity Techniques. In IEEE Network, vol. 29, no. 1, pp. 42–48, Jan.-Feb. 2015.

[21] H.-M. Wang, K.-W. Huang, and T. A. Tsiftsis. Multiple Antennas Secure Transmission under Pilot Spoofing and Jamming Attack. In IEEE Journal on Selected Areas in Communications, vol. 36, no. 4, 860–876, April 2018.

[22] Y. Liu, Z. Qin, M. Elkashlan, Y. Gao, and L. Hanzo. Enhancing the Physical Layer Security of Non-Orthogonal Multiple Access in Large-Scale Network. In IEEE Trans. Wireless Commun., vol. 16, pp. 1656–1672, March 2017.

[23] B. Reynders, W. Meert and S. Pollin. Range and coexistence analysis of long range unlicensed communication, In *Proceedings of 23rd International Conference on Telecommunications (ICT)*, Thessaloniki, 1–6, 2016.

[24] De Poorter, E., Hoebeke, J., Strobbe, M. et al. Sub-GHz LPWAN Network Coexistence, Management and Virtualization: An Overview and Open Research Challenges. Wireless Pers Commun, 95: 187, 2017.

[25] B. Vejlgaard, M. Lauridsen, H. Nguyen, I. Z. Kovacs, P. Mogensen and M. Sorensen. Interference Impact on Coverage and Capacity for Low Power Wide Area IoT Networks. In *proceedings of 2017 IEEE Wireless Communications and Networking Conference (WCNC)*, San Francisco, USA, 1–6, 2017.

[26] C. Liang and F. Yu. Wireless network virtualization: A survey, some research issues and challenges. *In Communications Surveys Tutorials, IEEE*, 17(1):358–380, 2015.

[27] F. Granelli, A. Gebremariam, M. Usman, F. Cugini, V. Stamati, M. Alitska and P. Chatzimisios. Software defined and virtualized wireless access in future wireless networks: scenarios and standards", *IEEE Communications Magazine*, 53(6), 2015.

[28] M. Yang, Y. Jin, L. Zeng, X. Wu, and A. Vasilakos. Software-Defined and Virtualized Future Mobile and Wireless Networks: A Survey, *Mobile Netw Appl*, 20(4), 2015.

[29] T. Cooklev, V. Poulkov, D. Bennett and K. Tonchev. Enabling RF data analytics services and applications via cloudification, In *Proceedings of IEEE Aerospace and Electronic Systems Magazine*, 33(5–6): 44–55, 2018.

[30] S. Boubiche, D. E. Boubiche, A. Bilami and H. Toral-Cruz. Big Data Challenges and Data Aggregation Strategies in Wireless Sensor Networks. *IEEE Access*, 6:20558-20571, 2018.

[31] A. L' Heureux, K. Grolinger and M. A. Capretz. Machine learning with big data: challenges and approaches. *IEEE Access*, 5:7776–7797, 2017.

[32] P. Baltiiski, I. Iliev, B. Kehaiov, V. Poulkov and T. Cooklev. Longterm spectrum monitoring with big data analysis and machine learning for cloud-based radio access networks. *Wireless Personal Communications*, 87(3):815–835, 2016.

[33] V. Navda, A. Bohra, S. Ganguly, and D. Rubenstein. Using channel hopping to increase 802.11 resilience to jamming attacks. In *Proceedings of IEEE INFOCOM 2007 – 26th IEEE International Conference on Computer Communications*, Barcelona: 2526–2530, 2007.

[34] Y. Liu, P. Ning, H. Dai, and A. Liu. Randomized differential DSSS: Jamming-resistant wireless broadcast communication. In IEEE INFOCOM: 1–9, 2010.

[35] M.J. Abbas, M. Awais and A.U. Haq. Comparative analysis of wideband communication techniques: Chirp spread spectrum and direct sequence spread spectrum. *In Processings of 2018 International Conference on Computing, Mathematics and Engineering Technologies (iCoMET)*, Sukkur: 1–6, 2018.

[36] T.T. Do, E. Björnson, and E.G. Larsson. Jamming resistant receivers for massive MIMO. In *Proceedings of 2017 IEEE International Conference on Acoustics, Speech and Signal Processing (ICASSP)*, New Orleans, LA: 3619–3623, 2017.

[37] H. Zeng, C. Cao, H. Li, and Q. Yan. Enabling jamming-resistant communications in wireless MIMO networks. In *Proceedings of 2017 IEEE Conference on Communications and Network Security (CNS)*, Las Vegas, NV: 1–9, 2017.

[38] B. DeBruhl and P. Tague. Digital filter design for jamming mitigation in 802.15.4 communication. In *Proceedings of 20th International Conference on Computer Communications and Networks (ICCCN)*, Maui, HI: 1–6, 2011.

[39] B. DeBruhl and P. Tague. Mitigation of periodic jamming in a spread spectrum system by adaptive filter selection. In *Proceedings of Int. Symp. Photon. Electromagn. Crystal Struct.*: 431–439, 2012.

[40] S.W. Arif, A. Coskun and I. Kale. Multi-stage complex notch filtering for interference detection and mitigation to improve the acquisition performance of GPS. In *Proceedings of 2018 14th Conference on Ph.D. Research in Microelectronics and Electronics (PRIME)*, Prague: 165–168, 2018.

[41] G. Iliev, Z. Nikolova, G. Stoyanov, and K. Egiazarian. Efficient design of adaptive complex narrowband IIR filters. In *Proceedings of XII European Signal Processing Conference, EUSIPCO 2004*, Vienna, Austria, III: 1597–1600, Sept. 2004.

[42] Z. Valkova-Jarvis, D. Mihaylova, A. Mihovska, and G. Iliev. Adaptive complex filtering for narrowband jamming mitigation in resource-constrained wireless networks, submitted.

4

Data Analytics for Securing the Smart Grid

Sarmistha De Dutta[1] and Ramjee Prasad[2]

CTIF Global Capsule, Aarhus University, Herning, Denmark
[1]Member, IEEE
[2]Fellow, IEEE
E-mail: sd56@columbia.edu; ramjee@btech.au.dk

The Smart Grid is the next-generation electrical power system consisting of a network of Internet of Things (IoT) devices and integrates two-way M2M communications with electricity flows for the efficient generation, delivery, and consumption of electrical energy. The integration of the two-way communication network, smart grid components, and the data they generate is a demonstration of the Communications, Navigation, Sensing, and Services (CONASENSE) model. However, along with the benefits of this evolving technology, the smart grid components, and the vast amount of 'big data' they generate, are susceptible to cyber attacks. The smart grid infrastructure can be protected using data analytics, specifically security analytics. The security analytics tools collect the big data generated by the smart grid segments and then analyze it to detect and proactively alert on potential cyber attacks. As more of the big data generated by the smart grid is being increasingly managed and processed by the cloud, the need for effective cloud security analytics is becoming more evident.

4.1 Introduction

The Internet of Things (IoT) refers to the connection of a wide variety of devices, with computing and communication ability, through a network. An IoT-enabled device may be referred to as a "smart" device, and depending

Security within CONASENSE Paragon, 59–74.

on its functionality, may have the ability to gather, analyze, and correlate the data generated by these smart devices. All IoT devices consist of three common components: hardware, network, and software. Advancements in data analytics allow the vast amounts of data generated by these smart devices to be efficiently analyzed. The IoT and its connected smart devices are being deployed in a wide range of verticals from agriculture, manufacturing, and transportation to healthcare and energy. By 2025 it is estimated that about 25–50 billion devices will be connected to the Internet, [1].

An increasingly popular application of IoT technology is in the energy sector, namely the smart grid. In the smart grid infrastructure, a system of devices such as sensors, smart appliances, or intelligent electronic devices (IEDs) are strategically installed on the M2M communications network and transmit and receive data which is then collected, analyzed, and accessed through a gateway such as the Internet. The smart grid allows utility companies to integrate two-way M2M communication and electricity flows. Sensors and advanced control mechanisms are used to capture then analyze real-time data related to power generation, delivery, and consumer usage. Utility providers use big data analytics to gain visibility into events across smart grid networks. This information allows utility companies to manage energy supply and demand, as well as customer billing, more efficiently, [1].

However, along with the undeniable benefits of the IoT, challenges in securing these enormous amounts of data have also arisen. Just like any system, the connectivity of IoT devices and networks are susceptible to security risks. Unauthorized agents may gain access and use these devices for malicious purposes. Cyber threats are growing increasingly complex and underlie the importance of securing these IoT products, both data and systems, through adequate security controls. Data analytics, and more specifically security analytics, which correlates database records, metadata, and pattern recognition, and machine learning (ML) algorithms is being used to detect, mitigate, and prevent cyber attacks on the grid.

Cloud-based architecture with its enormous data centers, computational power, and vast storage capabilities is another increasingly popular option for managing the big data generated by the smart grid. Cloud technologies provide flexibility, redundancy, and data backup across multiple locations, while reducing Capital Expenses (CAPEX) and Operating Expenses (OPEX) for utilities while also providing better data fault tolerance. The data generated by the smart grid is defined as a time series or data steam requiring specialized data processing. It is also important to note that this data, once stored in the cloud, is no longer under the control of the utility companies, and will need to adhere to defined standards and regulations that ensure the data is kept secure.

A cyber attack is an intangible threat that is difficult to predict. As more of the data generated by the smart grid is collected, stored, processed, and analyzed in the cloud, it increases the probability of a cyber-related attack. This continuous security data growth is also responsible in the migration of Security Information and Event Management (SIEM) systems to the public cloud infrastructure, with cloud-based SIEM solutions becoming increasingly popular. Cloud security analytics offers different methods to protect the data generated by the different segments of the smart grid and harden its security against cyber attacks.

The following section discusses the integration of the two-way communication network with smart meters, sensors, SCADA control systems, applications, services, as well as the data they generate and shows how the next-generation smart grid is an example of the integrated Communications, Navigation, Sensing, and Services (CONASENSE) paradigm, [2]. The later sections discuss about security issues facing the smart grid and ways to address them.

4.2 Smart Grid and Cyber Attacks on Energy Sector

The smart grid is the next-generation electrical power system that integrates Operations Technology (OT) with Information Technology (IT). The OT (computers/network for power delivery) platform allows utilities to manage energy resources, plan for emergencies, and maintain the grid load based on data received from among other sources, such as the Supervisory Control and Data Acquisition (SCADA) system. On the other hand, the IT (computers/network for business processes) platform is used by utilities to manage their business and administrative processes. This grid infrastructure utilizes increased use of telecommunications and digital technology for an automated, decentralized, consumer-interactive but also environmentally sustainable generation, delivery, and consumption of electrical energy. The smart grid uses communication and information technology to automate the electric power transmission and distribution systems and improve the efficacy of electricity usage, [3].

A smart grid relies on IoT devices to facilitate communication between the energy grid and the energy consumer. The National Institute of Standards and Technology (NIST) Smart Grid Conceptual Model defines the smart grid as consisting of seven distinct domains:

1. Bulk Generation
2. Transmission
3. Distribution
4. Customers
5. Operations
6. Markets
7. Service Providers

The first four domains cover the generation, storage, delivery of electricity, and support bi-directional flow. The latter three domains manage this electricity flow and provide data or services to power utility companies and their customers.

Energy is generated in the Bulk Generation domain from both centralized, traditional sources as well as distributed energy resources (DER): renewable variable sources – solar, wind; renewable non-variable sources – hydro, biomass, geothermal; non-renewable and non-variable sources such as nuclear, coal and gas.

The Transmission domain transports high voltage electricity over long distances. It may also store DER and generate electricity. A transmission electrical substation uses transformers to step up or step down voltage across the power grid.

In the Distribution domain, electricity is distributed to utility customers (home, commercial, industrial) who are connected to the electric distribution network through intelligent IoT devices such as smart meters. The Advanced Metering Infrastructure (AMI) is an integrated system of smart meters, meter data management systems, and gateways that allows communication between the customer and utility company.

Through the Operations domain, information such as energy consumption data collected from the consumer is further processed and provides key business intelligence. Integrating the AMI with the traditional distribution infrastructure has led to a more efficient power grid.

The sources of data generated by this intelligent smart grid are as follows:

- AMI (Smart Meters)
- Distribution automation (Grid)
- Third-party Data (Data external to Grid)
- Asset Management (Firmware for smart device or Operating System)

The smart grid is a much more complex system compared to its predecessor, the legacy power grid. The very components that are responsible for its 'intelligence' also make it more vulnerable to cyber threats. Vulnerabilities in the Smart Grid range from access to, and manipulation of, smart devices such as IEDs with malicious intent, proliferation of botnet malware, advanced persistent threats (APT), Denial of Service (DoS) attacks and Distributed Denial of Service (DDoS), smart meters hacked by external attackers or even by consumers themselves to steal power, modified communication channels by consumers to reduce their actual usage, deliberately modified sensor and SCADA systems data, to the unauthorized modification of routing in wireless networks and these can all disrupt grid operations. Cyber attacks against the energy sector have increased in recent years as shown in the Table 4.1 below, [3–5].

The smart grid is a vital and critical target that needs to be protected against cyber threats attacks. Each asset of the smart grid (i.e., smart meters, substation, etc.), and their communication channels, is a potential target for a cyber attack. These smart grid assets each have their own specific security requirements depending on its function. For example, the security controls needed to secure a substation will be different from those needed for the AMI. Whatever the differences, an attack on any asset may jeopardize the entire grid security, one that may have a cascade effect leading to a complete system blackout.

Table 4.1   Recent cyber attacks on energy industry

2015	2016	2017
• June – SCADA system data for hydroelectric generator exposed on Dark Web • December – SCADA system of Ukrainian electricity distributors attacked by BlackEnergy malware, New York dam infiltrated, and network breach of US natural gas and electricity company	• January – Ransomware email sent to Israeli Electricity Authority • March – Water utility's SCADA system controlling water flow and chemical level is hacked • April – Ransomware phishing on a Michigan Electric and Water utility, Malware found in Bavarian nuclear power plant	• June – Malware 'Industroyer' used in the cyberattack on Ukraine's power grid • December – Malware attack 'Triton' on Middle Eastern oil and gas petrochemical facility.

So, it is imperative to protect the entire end-to-end architecture of the smart grid by combining standard cyber security technologies such as antiviruses, firewalls, intrusion protection systems, with more advanced technologies such as security information and event management (SIEM), application whitelisting, and security features at the processor level, among others.

Cyber security solutions protect the smart grid against cyber threats by using comprehensive, real-time threat intelligence. Security information and event management (SIEM) systems collect and aggregate information from devices, networks, and applications. They analyze this internal and external data and use security events to identify risks and threats to the smart grid. Along with traditional malware protection technologies such as anti-virus, application whitelisting technologies ensure that only authorized files are executed. Hardware-assisted security is another option to make systems more robust and in the case of a cyber attack, reduces the time to return to normalcy, [6].

Some of the solutions that can be used to secure the smart grid against attacks are given in Table 4.2 below:

Managed Endpoint Detection and Response (EDR) using EDR tools provides a proactive security strategy to protecting the smart grid between endpoints with its ability to detect, identify, analyze, and respond to threats and intrusions. Anti-malware software may also be deployed at the smart grid endpoints to block against sophisticated malware from infiltrating the grid infrastructure. Threat intelligence provides context and actionable insight into an attack. Since cyber threats are constantly evolving, sharing this information can provide utility companies with an invaluable weapon in protecting the grid against cyber breaches. The smart grid communications network can be secured using integrated solutions for managing threat detection, forensic investigation, and the appropriate response. Another option for securing the network infrastructure is using a firewall. A firewall filters packets and only those packets that satisfy the user-defined rules are permitted to traverse the network. However, since firewalls examine packets at the lower Network layer of the OSI model, they may not be effective in detecting malicious code at the higher Application layer of the OSI model. Another defense mechanism for the network is the sandbox, an isolated environment on a network used to execute any potentially suspicious malware without compromising the security of the entire grid network.

As an alternative to firewalls, certain Intrusion Detection Systems (IDSs) and Active Denial System (ADSs) may be used to detect abnormal behavior

Table 4.2 Security solutions for smart grid cyber infrastructure protection

Endpoint Security	Threat Intelligence	Network Security	Mobile Security	Data and Applications	Identity and Access	Cloud Security	Fraud
Endpoint Detection and Response	Threat Sharing	Firewalls, Sandboxes	Transaction Protection	Data Monitoring and Data Access Control	Identity Management	Cloud Access Security Broker	Fraud Protection
Endpoint Patching and Management	IP Reputation	Forensics and Threat Management	Device Management	Application Scanning	User Access Roles and Privileges	Workload Protection	Criminal Detection
Malware Protection	Indicators of Compromise	Network Visibility, Segmentation, and Virtual Patching	Content Security	Application Security Management	Access Management		

in smart grid network infrastructure. The smart grid consists of a physical system and a cyber system, each supported by their own specific types of IDSs: These IDSs can be deployed to detect intrusive behavior in the smart grid. The utility operator can detect, control, and mitigate the fallout from a cyber attack if it receives an alert from an IDS alarm. Along with IDSs, Intrusion Detection and Prevention Systems (IDPSs) can also be applied to mitigate and deter a cyber attack. They typically are able to respond faster to a cyber attack on the smart grid, both before and after the actual occurrence.

The following types of IDSs are available and deployed based on the system component:

- Network-Based IDSs – The network-based IDS (NIDS) examines the packet header and payload of all packets traversing the network for anomalous behavior. Since each network packet has a unique format based on the communication protocol used, any anomalies can be identified by comparing any atypical packets against predefined rules.
- Host-Based IDSs – The host-based IDS (HIDS) identifies anomalies in measurements and the behavior of physical devices using rules that are user-defined to detect any behavioral irregularities.

A common method for detection of security breaches is the use of blacklists and whitelists. A common example of securing by means of blacklisting is the use of antivirus software. A virus is detected by comparing its signature against entries in a database. If the signature matches, the virus will be quarantined and may be sent to a sandbox for further analysis, or completely deleted. Whitelists meanwhile only allow access to authorized users, and software applications. To ensure protection against constantly evolving cyber threats, the rules used for both blacklists and whitelists must be constantly monitored and kept up-to-date.

However, detection of cyber attacks is only one part of keeping the smart grid secure. Once a threat is detected and an alert is triggered by the IDSs or IDPSs, its mitigation is the next phase in the defense cycle. The mitigation techniques will address both the computing infrastructure (i.e., computers, software applications, communications networks) and physical infrastructure (e.g., power system) of the grid, [7].

The security of the network can also be bolstered by splitting the network into smaller networks using the concept of network segmentation. Network segmentation will reduce the number of systems on the same network segment, thus limiting exposure to malicious activities. Secure data-centric access control also helps control access to applications and data, preventing security and privacy violations. Identity Access Management (IAM) and Privileged Access Management (PAM) also prevent un-authorized users from accessing the smart grid and the data it generates.

For a cloud-based smart grid, the protection of data, prevention of cyber threats, and policy violations can be addressed using a Cloud access Security Broker (CASB). A CASB is a centralized repository placed between the cloud service providers and consumers, managing policy and governance for users and devices whenever cloud-based resources are being used.

4.3 Data Analytics Platforms for the Smart Grid

With the continual emergence of new attackers (criminals, nation state actors, non-state actors) on smart grids it has become more important to detect cyber attacks on smart grid systems using an advanced security analytics platform using public and private threat intelligence with following components, [8]:

- Big Data – Collect data from different sources and store it for further analysis
- Analytics – Alert and Report, Investigate and Analyze, Visualize, and Respond
- Governance – Compliance, Incident Management, and Remediation

Data analytics, or data mining, is the process to compute and determine possible interrelationships between variables using databases, metadata and statistics, pattern recognition, and ML algorithms, etc. These related techniques are called security analytics and have a key role in processing big data, the basis for discovering invaluable information that will ultimately help in the decision-making for protecting the smart grid, [9].

The smart grid supports the following kinds of data analytics as shown in Figure 4.1, [10]:

- Signal Analytics – Based on signal processing (e.g., Sensor signals, substation waveforms, and line sensor waveforms, non-usage meter data)
- Event Analytics – Focused on events. (e.g., Detection, Classification, Filtering, and Correlation)
- State Analytics – Provides insight into the state of the grid. (e.g., Real time electrical state, real time grid topology, and system identification)

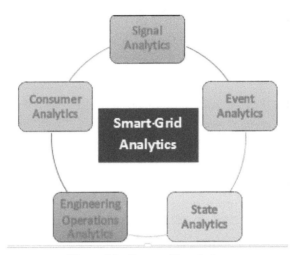

Figure 4.1 Smart grid analytics.

- Engineering Operations Analytics – Provides information about grid operations such as operational effectiveness, system performance, as well as load trends and forecasting.
- Consumer Analytics – Provides situational awareness based on customer data. (e.g., Demand profiles, demand response, diversion analytics, and customer segmentation)

The operational aspect of the smart grid may be defined by models such as descriptive, diagnostic, predictive, and prescriptive (refer to Figure 4.2) that are combinations of the above-mentioned analytics classes. The descriptive model is used to describe customer behavior and provide insight into their usage. The diagnostic model is used to understand the behavioral patterns of particular users and analyze their decisions. These models are useful to make predictive models that forecast the decisions that customers may make in the future. Lastly, the prescriptive model is the highest level of smart grid analytics since it affects important aspects such as marketing, engagement strategies and actual decision-making.

Depending on the type of analytics, the big data must be identified, stored, and then processed for further analysis. Apache Hadoop is a popular open-source project supports distributed storage and processing of large data sets. Some other big data processing engines include Storm, Spark, and Flink, [10].

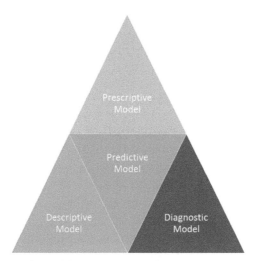

Figure 4.2 Analytic models for smart grid.

4.3.1 Consumer Analytics

As energy consumers-producers aka "prosumers" are connected to the decentralized electric distribution network through the Advanced Metering Infrastructure (AMI), the analysis of customer-generated data is important for utilities to operate efficiently and fulfil the full potential of the smart grid. Utility companies use Demand Response (DR) programs to gain real-time data about the electricity demands and help them balance the grid load while simultaneously helping the consumer manage their energy needs cost effectively. Consumer analytics on big data can be performed using the following processing modes, [10]:

1. Batch processing – Used to process data within a specific period of time when there is no constraint on response time. Hadoop, an open source software utility that performs data processing and storage for big data applications is a suitable option for the geographically distributed smart grid.
2. Real time stream processing – Used for real-time applications that require a response with very low latency. (e.g., Splunk, Storm)
3. Hybrid processing – Handles both batch and stream processing (e.g., Flink, Spark)

4.4 Security Analytics for the Smart Grid

A secure and high-performance data analytics platform is crucial to protect the smart grid against cyber attacks.

The smart grid data flows from power plants and physical devices such as smart meters, sensors, etc. The data from power plants is streamed to control systems and used for monitoring the status of the grid in real-time or storing the data for future forecasting of power consumption trends. The physical devices meanwhile generate data with a timestamp, then transmit the data through the network either as a time series or a data stream. A data stream is defined as a sequence of time-stamped records. Each record comprises of a key-value pair where the key represents the reading and the value is the corresponding data reading, [11, 12].

As we mentioned earlier, the very online connectivity that makes the smart grid "smarter" also makes it more susceptible to cyber-attacks. Security measures may be executed at the generation side, transmission side, or consumer side. With innovation in global smart grids, including the expansion of the Internet of Things (IoT), and the evolving cyber attacks on this infrastructure, the smart grid cyber security market is estimated to grow to grow at a rate of 10.01% during the period 2017 – 2021 with a market value of USD +6.00 Billion by 2023.

The growth in the global smart grid security market, by function is shown below:

The security analytics solutions for effectively hardening the security of the smart grid should include the following features, [5]:

- Incident Response – The process to quickly identify, address, manage, and determine the root cause of a cyber attack.
- User Behavior Analysis – Identify patterns in human behavior, then use algorithms and statistical analysis to detect anomalous behavior that may indicate possible threats, from those patterns.
- Threat Hunting – Proactively search for malware or cyber attackers who may have already infiltrated the smart grid infrastructure, investigate, and address any resulting security issues.
- Cognitive Security – Use Artificial Intelligence (AI) that has been designed based on human thought processes to detect potential security threats.
- Vulnerability Management – The process of identifying security vulnerabilities and then remediating and mitigating them to prevent unauthorized access to the smart grid systems and data.

- Threat and Anomaly Detection – The process to identify unusual patterns that differ significantly from the anticipated behavior, indicating a cyber threat.

These security analytics tools allow the smart grid to collect the data, then analyze it for detecting and proactive alerting of potential cyber attacks.

4.5 Cloud Computing and Security Analytics for Smart Grid

With the emergence of cloud computing, another option for managing and processing the enormous data sets generated from the smart grid is now available. The cloud network offers private, public, community, and hybrid deployment models. Public cloud providers such as Amazon Elastic Compute Cloud (Amazon EC2), Google Cloud Platform, Microsoft Azure, IBM Cloud, and Oracle Cloud are shown in Figure 4.3. Some popular Private Cloud providers are VMWare. Oracle VM, OpenStack, Microsoft Hyper-V, and CITRIX Xen Server.

A key component of the cloud-based smart grid infrastructure is the database (DB) system. The database management system (DBMS) and data mining (i.e., data analysis) features, stores the data logically and then retrieves, processes, and analyzes it, either in real-time or retrospectively. Relational databases (e.g., Oracle, SAP Sybase, MySQL, PostgreSQL, etc.)

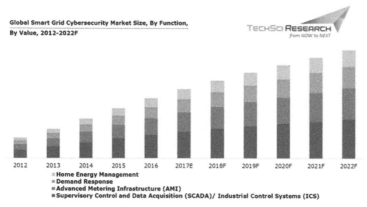

Figure 4.3   Growth in global smart grid cybersecurity market.

Source: TechSci Research.

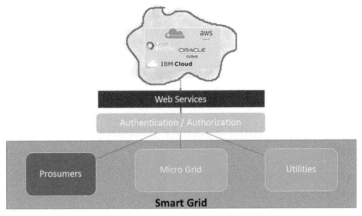

Figure 4.4   Smart grid cloud architecture.

are widely used in the RDBMS and data is retrieved using Structured Query Language (SQL). Non-relational database systems, known as NoSQL, are increasingly being used for storing and processing the big datasets generated by the smart grid. The NoSQL databases (e.g., MongoDB, BaseX, Apache Cassandra, etc.) partition the data and store the partitions across multiple servers' thus providing data redundancy in case of server failure. Along with cloud computing, the cloud database is also another option to store the terabytes or petabytes of smart grid data. The cloud database may either be a virtual machine instance or in the form of a Database as a Service (DBaaS), [13].

In conjunction with these databases, and as mentioned in an earlier section, the open source software framework Hadoop provides reliable and scalable distributed computing, making it suitable for the processing of the smart grid big data. Hadoop uses the following concepts:

- Hadoop Distributed File System (HDFS) – A distributed file system that provide high-throughput access to application data by distributing the processing of large data sets across computer clusters.
- Hadoop MapReduce – Distributes a large data set across multiple computers for parallel processing.
- Hadoop YARN – Manages job scheduling and cluster resources

Along with cloud computing, the cloud database is also another option to store the terabytes or petabytes of smart grid data. The cloud database may be a virtual machine instance or a Database as a Service (DBaaS).

Cloud computing provides the following types of applications [14]:

- Infrastructure as a Service (IaaS) – Infrastructure service model provides a hardware platform on an on-demand basis. It includes storage, virtual machines and is used for load balancing in cloud computing.
- Platform as a Service (PaaS) – Delivers programming models to IaaS through the cloud and allows users to execute programs.
- Software as a Service (SaaS) – Supports all applications in the cloud environment via web browsers.
- Database as a Service (DBaaS) – Installs and maintains the cloud database that stores the data.

The advantages of using cloud computing for the smart grid are as follows:

- Elastic Nature – Supports elastic nature of storage and memory devices, based on the requirements of the end users.
- Shared Architecture – Supports shared architecture allowing information to be easily shared for cost effectiveness.
- Metering Architecture – Offers metering infrastructure to consumers allowing them to easily control and monitor their energy usage.
- Internet Services – Cloud computing supports the existing network infrastructure.

Cloud analytics provides access to data analytics tools, through a public or private cloud, using a subscription-based or pay-per-use pricing model. Cloud analytics is any data analytics or business intelligence (BI) process used in collaboration with a cloud provider. Since existing cloud-based security applications can be used for securing the smart grid, this makes it a cost-effective solution [14, 15].

Summary

The integration of cloud computing and the smart grid can monitor, process, and manage the big data generated from the different elements of the smart grid while providing robust security controls compared to the existing IP-based security controls.

The ever-growing demand for an uninterrupted and efficient power supply will result in the increasing need for smart grid deployment. The need to protect the smart grid against cyber-attacks, will only fuel the need for end-to-end security analytics solutions. Vendors such as Cisco, Intel (McAfee), Siemens, Symantec, IBM, etc. offer out-of-the-box security analytics solutions that can be customized based on the smart grid infrastructure.

These security analytic solutions convert the raw data generated from various smart grid data sources into actionable intelligence that correlate activities and alerts. This information can then be used to identify security threats and trigger alarms, thus protecting the smart grid from a cyber attack.

References

[1] US GAO Center for Science, Technology, and Engineering (Technology Assessment report – Internet of Things).

[2] Prof., L.P. Ligthart, Prof., Ramji Prasad, "CONASENSE; Communicat ions, Navigation, Sensing and Services" River Publishers Series in Communications.

[3] Sarmistha De Dutta, Prof. Ramji Prasad, "Cybersecurity for M2M Smart Grid" Global Wireless Summit (GWS) 2015, Hyderabad, India.

[4] "Internet Security Report," Symantec, Volume 22, April 2017.

[5] Joe DiAdamo,"Smart Grid Security," IBM, 2018.

[6] "Smart Grid Cyber Security," Intel, Alstom, McAfee, 2013.

[7] Chih-Che Sun, Chen-Ching Liu et al., "Cyber-Physical System Security of a Power Grid: State-of-the-Art," Electronics 2016.

[8] Dr. Robert W. Griffin, "Security Analytics and Smart Grid Security," RSA-EMC, 2014.

[9] https://energyinformatics.springeropen.com/articles/10.1186/s42162-01 8-0007-5.

[10] https://journalofbigdata.springeropen.com/articles/10.1186/s40537-017-0070-y.

[11] https://www.businesswire.com/news/home/20181010005893/en/Global-Smart-Grid-Cyber-Security-Market-2017-2021.

[12] Adela Beres, Bela Genge et al., A Brief Survey on Smart Grid data analysis in the Cloud," 8th International Conference Interdisciplinarity in Engineering (INTER-ENG), Tirgu-Mures, Romania, INTER-ENG 2014.

[13] Zeyar Aung, "Database Systems for the Smart Grid."

[14] Samaresh Bera, Sudip Misra et al., "Cloud Computing Applications for Smart Grid: A Survey," IEEE Transactions on Parallel and Distributed Systems, 2014.

[15] Adela Beres, Bela Genge et al., A Brief Survey on Smart Grid data analysis in the Cloud," 8th International Conference Interdisciplinarity in Engineering (INTER-ENG), Tirgu-Mures, Romania, INTER-ENG 2014.

5

EM-Based Drone-Detection/Identification for Safety Purposes

Ernestina Cianca, Simone Di Domenico, Mauro De Sanctis and Tommaso Rossi

Department of Electronic Engineering, University of Rome Tor Vergata, Italy
E-mail: cianca@ing.uniroma2.it; simone.didomenico@uniroma2.it; mauro.de.sanctis@uniroma2.it; tommaso.rossi@uniroma2.it

Drone surely could be effectively used to improve the safety in several applications (e.g., transport applications, surveillance). On the other hand, they also represent a threat to safety and security. Therefore, both for civil and military applications, it is crucial to develop cost-effective detection systems to warn of the presence of drones. In many military applications, it is also important to be able to identify the drone at some extent (the model, dimension etc.). This Chapter reviews the main approaches proposed for drone detection/identification, mainly focusing on the class of techniques based on the analysis of some Electromagnetic (EM) signature, which includes:

1. Detection based on the analysis of the signature of Radio Frequency (RF) signals intentionally emitted by the drone.
2. Detection based on the analysis of the signature of more general EM signals not intentionally emitted by the drone.
3. Detection based on the analysis of the signature of RF signals reflected by the drone and transmitted by sources of opportunity.

Some original results are presented and open challenges highlighted.

Security within CONASENSE Paragon, 75–84.

Introduction

The drone market is the fastest growing in aerospace, both in the public sector at large (safety, security, environment monitoring, etc) and in the private sector (farming, infrastructure, delivery, inspection, broadcasting, leisure etc.). Even if they could be effectively used to improve the safety in several applications (e.g., transport applications, surveillance), they also represent a threat to safety and security. Just as examples: in 2014, they have been used to deliver drugs into some jails in the South of Carolina [1]; they have been used to hack WiFi networks inside company buildings to corrupt the correct operations of devices into the company [2]; more in general, drones are increasingly flying in sensitive airspace where their presence may cause harm, such as near airports, forest fires, large crowded events, secure buildings, and even jails.

The presence of drones has interfered with and grounded aircraft fighting forest fires [3]. Drone crashes have also disrupted sporting events such as the US Open tennis tournament as well as a World Cup skiing race [4, 5].

A variety of approaches have been explored to interdict drones [6, 7]. However these interdiction strategies typically presume that the presence of the drone has already been detected.

Therefore, both for civil and military applications, it is crucial to develop cost-effective detection systems to warn of the presence of drones, and in some cases, of "authorized drone" or not "authorized drones". In many military applications, it is also important to be able to identify the drone at some extent (the model, dimension etc.).

In this framework, several works have been done recently. Each of them is characterized by different advantages and disadvantages [8]. In the rest of this Chapter, different detection approaches are presented, outlining pro and cons and reporting recent results. This Chapter focuses on the detection techniques based on the analysis of EM signatures. Some original results are presented.

5.1 Drone Detection Approaches

Current solutions for drone detection include: visual techniques [9], active radars [10], detection of acoustic signature [11] and a wide range of techniques based on the analysis of some electromagnetic (EM) signature [12].

Active radar approaches requires LOS conditions, which could be not applicable in case of urban areas. Moreover, RCS should be sufficiently large, or alternatively more power and bulky transmitter must be used. Therefore, the small size of most of the drones to be detected poses great detection limitations [10].

Audio-based approaches can be confused by other sounds in noisy environments, has limited range, and cannot detect drones that employ noise canceling techniques. Solutions based on video detection require the presence of either distributed camera equipment or 360° video recording devices. In both cases, the cost of the infrastructure could be high and they suffer from poor visibility problems. Thermal and IR imaging cameras for long distance are prohibitively expensive and have limited coverage.

In this Chapter we focus on the class of techniques based on the analysis of some EM signature, which includes:

1. Detection based on the analysis of the signature of RF signals intentionally emitted by the drone.
2. Detection based on the analysis of the signature of more general EM signals not intentionally emitted by the drone.
3. Detection based on the analysis of the signature of RF signals reflected by the drone and transmitted by sources of opportunity.

In all the mentioned methods, the needed detection infrastructure could be installed on the ground or on another drone explicitly used for detection [13].

Signature of RF signals intentionally emitted

These techniques assume that the drone intentionally transmit an RF signal, for instance, to communicate with the controller. Most of the drones usually communicate with their controllers frequently around 30 times per second to update its status and to receive the commands from controller [12, 14, 26]. Moreover, with respect to the communication channel between an access point (AP) and mobile devices, which usually exchange beacons at every 100 ms (10 Hz), the drone controller requires higher frequency of communication to control the drone precisely. Therefore, wireless samples can be collected at higher rates than in case of typical WiFi transmission, thus improving the detection system performance (more information per unit time). This type of detection systems requires some level of knowledge of the transmitted signal.

Some commercially available drones uses proprietary communication protocols, whose characteristics are not easily available. However, most of them have the following characteristics [14, 15]:

- use of ISM bands
- the communication protocol between drone and controller can be programmed and personalized.
- The modulation order can also be selected.

- They use SS techniques, either FH or DSSS, or a combination of them.
- In some cases, drones do not transmit unless specifically triggered.

In some of them, the ones most commonly used for leisure purposes, the connection is established using a WiFi link. The idea of this type of approaches is that the received signal contains specific "signatures" that are impressed by the drone. For instance, the RF propagation channel is affected by the fact that the transmitter moves. This movement causes a Doppler shift in the received signal. Therefore, by studying the Doppler spectrum of the received signal, it should be possible to understand if the transmitter is on-board a moving means. In [16], the Doppler spectrum of a WiFi signal is analyzed to measure the time variability of the channel, which is assumed to be correlated to the number of people in the monitored room. The signal is received by a WiFi receiver and uses only the overhead of the packet, which contains the same information and hence, the fluctuation of the received signal can be associated to the fluctuation of the propagation channel. The same approach could be used in a drone detection system.

It must be noted that, especially in urban areas, the WiFi signal could come by other sources, different than drones. Therefore, it is important that the detection method is able to distinguish between the signal transmitted from a drone or from other sources.

To prove the feasibility of this approach, experimental tests have been done. The experimental set-up consider a drone with a WiFi transmitter, one person carrying a smartphone with active WiFi transmitter, and a WiFi card Intel WiFi Link 5300 with 3 antennas which acts as receiver. CSI are extracted from the received packets by using a customized firmware and an open source Linux wireless driver for the Intel 5300 WiFi card. Then, collected CSI are processed as described in [16] to calculate features related to the propagation channel and in particular, to the Doppler spectrum. Figure 5.1 shows the scatter plot for two features, the slope and the 3rd order moment of the Doppler Spectrum. Table 5.1 shows the results in terms of confusion matrix. The accuracy of the system is on the average 0,9323. The drone is confused with a sitting person only for the 8% of cases. Results are promising but more extensive experiments should be done to understand the coverage of the detection methods and also the capability to detect multiple drones.

Other methods are described in [17], where the transmitted RF signal is studied to identified signatures related to other drone-specific features such as two key inherent types of movements of the drone's body: body shifting caused by the spinning propellers; body vibration due to navigation and

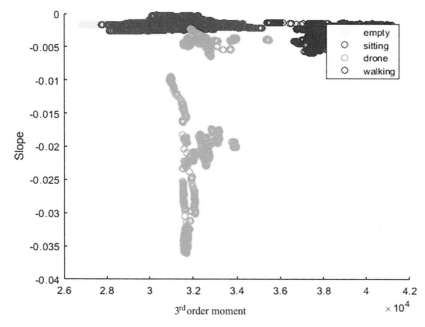

Figure 5.1 Scatter plot showing two features calculated on the Doppler Spectrum estimated by the collected CSI.

Table 5.1 Confusion matrix

		Prediction			
		Empty	Sitting	Walking	Drone
TRUTH	Empty	0.8083	0.1917	0.0000	0.0000
	Sitting	0.0000	1.0000	0.0000	0.0000
	Walking	0.0000	0.0000	1.0000	0.0000
	Drone	0.0000	0.0791	0.0000	0.9209
					0.9323

environmental impact corrections. The controller needs to frequently react to compensate for unpredictable environmental phenomena, such as wind, inaccuracy of sensors and actuators, thus causing undesirable physical movement of the drone, which are leveraged as unique signatures. As in the previous example, in [17] the transmission link is based on WiFi. Authors demonstrate that this system can detect the physical signatures to uniquely identify an individual drone and effectively differentiate it from other mobile wireless devices at distances of hundreds of meters.

The main limitation of these approaches is the need of an intentional communication link and also the some level of knowledge of the communication protocol, which is not always possible.

Signature of RF signals reflected by the drone

In this Section, we do not consider the case in which the reflected signal is transmitted by an active radar but only signals transmitted by sources of opportunity which could be:

– Global Navigation Satellite Systems (GNSS) [18]
– Wi-Fi [19]
– Digital Video Broadcasting-Terrestrial (DVB-T(2)) [20, 21]
– Global System for Mobile communications (GSM) [22] and Long-Term Evolution (LTE) [23].

As a matter of fact, the reflected signal could be also the signal transmitted by the drone controller. In [12], the drone is detected based on the signature of the signal, transmitted by the controller and reflected from its propellers, which could be observed by off-the-shelf wireless receiver (i.e., wifi receiver).

The reflected signal is not continuous and its duty cycle depends on the rotation speed and size of the propeller, and the distance between the drone and receiver.

For example, if the propeller rotates with the speed around 7500 to 10500 RPM (as in Bebop ARDrone [24]), the signature of the drone in expected on the frequency band less than 200Hz. However, the reflection capability depends on the drone orientation and distance with respect to the wireless receiver.

Signature of more general EM signals not intentionally emitted by the drone

In case the drone does not transmit or it is not possible to know anything about the communication protocol, the only detection approach based on EM waves is the one based on not intentional EM emissions. Figures 5.2(a) and 5.2(b) show measurements of the electric field, for both polarization, in the range 200MHz-1 GHz. The measurements have been conducted in a semi-anecoic chamber, on the Phantom 3 UAV model produced by the chinese company

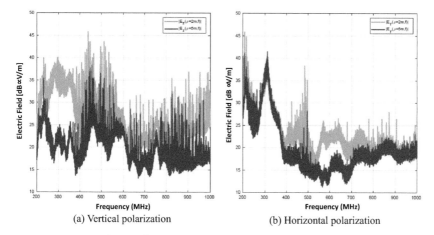

(a) Vertical polarization   (b) Horizontal polarization

Figure 5.2 Electric field measured on a Phantom 3.

DJI. Two distances of the drone from the detection system have been considered: 2m and 5m. From Figures 5.2(a) and 5.2(b), it can be observed that:

- The electric field is detectable, and there are distinguishable frequency peaks. It should be better understood if those peaks are correlated to some drone-specific feature and used to detect and in case, identify the drone. It is worth noting that other measurements, on other type of drones show a different "spectrum signature".
- Several frequency components show a high level of polarization. This level of polarization could represent an element of identification of the model of drone as the polarization is strongly dependent from the geometry of the radiating object.

Next step is the understanding of the correlation of the frequency peaks with some specific characteristic of the drone, for instance, if they could be correlated to the speed of the drone, or other typical movements of the drone. Some preliminary studies are on-going but mainly in military field and no much can be known.

As previously mentioned, the detection system could be placed on another "surveillance" drone. However, this approach seems not feasible in this case for several reasons: i) the surveillance drone would generate other EM emissions, which would result in interference; ii) the detection infrastructure could be too big and heavy. The previous measurements have been conducted in an interference-free environment. As a matter of fact, in a real scenarios much more EM interference is present and also detection distances should be bigger. Therefore, it is expected that such a detection system should use big antennas and costly infrastructure to be able to really "hear" these signatures in a real noisy environment. On the other hand, this could be the only approach is some applications, and especially for military or security critical applications, it would be important to further investigate its feasibility.

References

[1] https://www.nytimes.com/2015/04/23/us/drones-smuggle-contraband-over-prison-walls.html?_r=1

[2] https://www.usenix.org/legacy/event/woot11/tech/final_files/Reed.pdf

[3] P. McGreevy. Private drones are putting firefighters in 'immediate danger,' california fire official says. LA Times, August 18, 2015.

[4] D. Waldstein. Drone crash interrupts match. New York Times, September 3, 2015.

[5] CNN. Drone crashes onto piste, misses champion skier by inches. https://goo.gl/jtPdLh, 2015. [Accessed Nov 01, 2016].

[6] R. Vander Schaaf. What technologies or integrating concepts are needed for the US military to counter future missile threats looking out to 2040? PhD thesis, US Army, 2014.

[7] J.-S. Pleban et al. Hacking and securing the ar. Drone 2.0 quadcopter: investigations for improving the security of a toy. In SPIE Electronic Imaging, 2014.

[8] M. M. Azari, H. Sallouha, A. Chiumento, S. Rajendran, E. Vinogradov, and S. Pollin, "Key technologies and system tradeoffs for detection and localization of amateur drones," IEEE Communications Magazine, vol. 56, no. 1, pp. 51–57, Jan 2018.

[9] A. Rozantsev, S. Sinha, D. Dey, and P. Fua, "Flight dynamics-based recovery of a uav trajectory using ground cameras," in Conference on Computer Vision and Pattern Recognition (CVPR), July 2017.

[10] F. Hoffmann, M. Ritchie, F. Fioranelli, A. Charlish, and H. Griffiths, "Micro-Doppler based detection and tracking of UAV swith multistatic radar," in IEEE Radar Conference (RadarConf), May 2016, pp. 1–6.

[11] J. Kim, C. Park, J. Ahn, Y. Ko, J. Park, and J. C. Gallagher, "Real-time UAV sound detection and analysis system," in IEEE Sensors Applications Symposium (SAS), March 2017, pp. 1–5.

[12] P. Nguyen, M. Ravindranatha, A. Nguyen, R. Han, and T. Vu, "Investigating Cost-effective RF-based Detection of Drones," in Proceedings of the 2nd Workshop on Micro Aerial Vehicle Networks, Systems, and Applications for Civilian Use. ACM, June 2016, pp. 17–22.

[13] Vinogradov, Evgenii and Kovalev, Dmitry and Pollin, S. (2018). Simulation and Detection Performance Evaluation of a UAV-mounted Passive Radar. 10.1109/PIMRC.2018.8580940.

[14] S. Piskorski et al. Ar. drone developer guide. Parrot, sdk, 1, 2012.

[15] T. Andre et al. Application-driven design of aerial communication networks. Communications Magazine, IEEE, 52(5):129–137, 2014.

[16] S. Di Domenico, G. Pecoraro, E. Cianca, and M. De Sanctis. "Trained-once device free crowd counting and occupancy estimation using WiFi: A Doppler spectrum based approach." In: 2016 IEEE 12th International Conference on Wireless and Mobile Computing, Networking and Communications (WiMob). 2016, pp. 1–8. doi: 10.1109/WiMOB.2016.7763227.

[17] Nguyen, et al. (2017). Matthan: Drone Presence Detection by Identifying Physical Signatures in the Drone's RF Communication. 211-224. 10.1145/3081333.3081354.

[18] X. He, T. Zeng, and M. Cherniakov, "Signal detectability in SSBSAR with GNSS non-cooperative transmitter," IEE Proceedings Radar, Sonar and Navigation, vol. 152, no. 3, pp. 124–132, June 2005.

[19] P. Falcone, F. Colone, and P. Lombardo, "Potentialities and challenges of WiFi-based passive radar," IEEE Aerospace and Electronic Systems Magazine, vol. 27, no. 11, pp. 15–26, Nov. 2012.

[20] M. Conti, F. Berizzi, D. Petri, A. Capria, and M. Martorella, "High range resolution DVB-T Passive Radar," in The 7th European Radar Conference, Sept 2010, pp. 109–112.

[21] A. Barkhatov, E. Vorobev, and A. Konovalov, "Experimental results of DVB-T2 passive coherent location radar," in 2017 IEEE Conference of Russian Young Researchers in Electrical and Electronic Engineering, Feb 2017, pp. 1229–1232.

[22] P. Krysik, P. Samczynski, M. Malanowski, L. Maslikowski, and K. S. Kulpa, "Velocity measurement and traffic monitoring using a GSM passive radar demonstrator," IEEE Aerospace and Electronic Systems Magazine, vol. 27, no. 10, pp. 43–51, Oct

[23] G. Pecoraro, S. Di Domenico, E. Cianca and M. De Sanctis, "LTE signal fingerprinting localization based on CSI," *2017 IEEE 13th International Conference on Wireless and Mobile Computing, Networking and Communications (WiMob)*, Rome, 2017, pp. 1–8. doi: 10.1109/WiMOB.2017.8115803

[24] Julio. Parrot bebop quadcopter review: A (near) flawless drone with the skycontroller. http://tinyurl.com/h394ljk. FPV Drone Reviews, February 05, 2016.

6

UAV Communication Networks: Problems and Possible Solutions

Homayoun Nikookar

Netherlands Defence Academy, The Netherlands
E-mail: h.nikookar@mindef.nl

Recently the military and civilian applications of Unmanned Aerial Vehicles (UAVs) have increased remarkably. In the past UAVs were mainly used for surveillance or as practice-targets for anti-aircraft artillery. Nowadays they are used for coverage extension, in search and rescue operations, in hazardous site inspections, and in many more commercial applications. The individual UAV nodes in a network can transmit data to each other. The coverage area of the network could be large enough to have UAVs in the network that are not within Line-Of-Sight (LOS) of each other. UAV communication networks are usually called Flying Ad Hoc NETworks (FANETs). FANETs are still in the development stage and face a number of challenges. The purpose of this chapter is to identify those challenges, see which technologies have been put forward and examine whether those technologies can overcome the FANET challenges and what possible consequences they may have.

6.1 Introduction

6.1.1 UAV Communication Networks

Recently the military and civilian applications of Unmanned Aerial Vehicles (UAVs) have increased remarkably. In the past UAVs were mainly used for surveillance or as practice-targets for anti-aircraft artillery. Nowadays they are used among others, for coverage extension, search and Rescue operation, hazardous site inspection, and many more commercial applications. Furthermore, in recent years advances in artificial intelligence have allowed the development of the autonomous UAVs. The next step is communication of autonomous UAVs in order to perform tasks as a group. In 2017 1,180 autonomous UAVs performed a light show in China. All the movements were synchronized and the UAVs were programmed to automatically land if they were unable to achieve their objective [1]. The US navy dropped 103 autonomous UAVs from an F18 Hornet in mid-flight and had them perform manoeuvres [2]. Several possible civilian and military uses for communicating UAV networks have been proposed. Especially in the military applications they could overwhelm enemy air defences, carry out surveillance or jam enemy communications over a large area. Their biggest advantage would be their ability to handle heavy losses, due to their sheer size [3]. Further in this direction, UAV communication networks could also act as a backbone for combat ground networks and relay their data to headquarters [4].

The individual UAVs nodes in a network can transmit instructions and data to each other. The coverage area of the network could be large enough to have UAVs in the network that are not within Line-Of-Sight (LOS) of each other. UAV communication networks are usually called Flying Ad Hoc NETworks (FANETs) [5]. These flying networks are still in the development stage and face a number of challenges, before they can be fully fledged. The purpose of this chapter is to identify those challenges, see which technologies have been put forward and examine whether those technologies can overcome the challenges and what operational consequences do they carry with them.

6.1.2 Research Questions

Research question: Can current technologies for FANETs solve the problems FANETs face and what operational consequences do these technologies have?

The research question can be divided into the following sub-questions.

- Which major problems do FANETs networks currently face?
- Which technologies can be used to solve these issues and how do these technologies work?
- Are the technologies capable of solving the problems?
- What operational consequences will the technologies have?

6.2 Method

The chapter is based on survey of literature. In Section 6.3 the problems facing FANETs will be identified. In Section 6.4 for each type of problem a technology will be presented and explained. In Section 6.5 the effectiveness of the technologies is evaluated and their operational consequences discussed. In the conclusion the main research question will be answered.

6.3 Problems

The following four issues do NOT represent all the challenges that need to be overcome in the development of FANETs. They are, nevertheless, significant problems that require a solution, in order to make the concept of FANETs work.

6.3.1 Aircraft Altitude and Mobility

Differences in altitude between the UAVs in a communication network and the high mobility of UAVs are inherent characteristics of a UAV network. They also pose a significant problem for communication between the UAVs. The locations of UAVs can change rapidly, [5, 6]. Depending on the type of engines of the UAVs in the network and whether they are fixed-wing or rotary-wing, locations could change in the range of hundreds of meters within seconds. This influences the decision on which type of antenna to use. UAVs can either use omnidirectional antennas or directional antennas. Omnidirectional antennas can transmit in all directions but suffer from power loss over distance and susceptibility to interference and jamming. Directional antennas compensate power loss with a higher antenna gain, but they have to be steered towards the receiving antenna [7].

6.3.2 Changing Topology

The above mentioned mobility characteristics also influence the topology of the network. Individual UAVs can also be lost due to malfunctions or enemy fire. Connections between UAVs can, therefore, be lost frequently. In a FANET data between UAVs that are not within LOS should be possible through multi-hopping. The routes data can take will be lost often, because of frequent connection losses between UAVs. The FANET is then required to find new possible routes, in order to make communication possible again. The routing protocol used by the FANET should be capable of coping with these frequent changes [8].

6.3.3 Clock Synchronization

FANETs could use Frequency Hopping techniques. These techniques require time synchronization between all UAVs in the FANET. In a FANET, however, UAVs can enter or leave the network at a high rate. This implies that entries in the FANET also happen frequently. Each UAV that connects to the FANET needs to perform a clock synchronization. Otherwise, the UAV will not be able to listen to the right frequencies. FANETs that use frequency hopping require periodic beacons that provides timing information. With a high number of UAVs transmitting intermittently, contention could become a problem. Contention could prevent such a beacon from being received by UAVs [9].

6.3.4 Security

The security of a UAV network needs to be robust in order to prevent spoofing of UAVs. For example, in December 2011 Iran was able to spoof a US UAV and land it on one of its own airbases [10]. Such spoofing activities could seriously endanger a UAV network. To prevent the entry of 'rogue' UAVs that could spoof other UAV's, all entering UAVs in a FANET have to be authenticated. In wireless networks this is done using symmetric- or asymmetric key cryptosystems. A large FANET using a symmetric key cryptosystem would require a vast amount of keys, because for every pair of UAVs that could connect there needs to be a separate key to encrypt the data. The alternative, a public key cryptosystem, requires a public key infrastructure. This entails the need for either key distribution centres (KDC) or certification authorities (CA) [11]. FANETs that have so far been developed do not have a public key infrastructure.

6.4 Technologies

For all the problems elaborated on in Section 6.3, one or two technologies will be given that could solve the problem. There are more technologies being developed to overcome the challenges.

6.4.1 Aircraft Altitude and Mobility

To address the problem of differing altitudes and high mobilities, a MAC-protocol has been developed that can use both omnidirectional and directional antennas. The UAVs should have two directional and two omnidirectional antennas. Furthermore, one directional antenna will be placed on top of the UAV and one underneath [7]. This should provide 3-dimensional coverage. The directional antennas need to be steered towards the target UAV and require detailed information about the location of the target UAV and the Euler angles of its directional antenna. The UAVs should be equipped with GPS-receivers and Inertial Measurement Units (IMU). The MAC-protocol to acquire the information works as follows. If the UAV is not transmitting, it listens using its omnidirectional antennas to other UAVs. These UAVs will send a so-called heartbeat which contains the location of the UAV and the Euler angles. Before a UAV wants to transmit data to another UAV using its directional antennas, it sends a Clear To Send (CTS)/Request To Send (RTS). The CTS will contain the most up-to-date information about the location of the transmitting UAV. The target UAV will respond with an acknowledge (ACK) that includes its own parameters. Both the CTS/RTS and the ACK will be send using omnidirectional antennas. Now the directional antennas will align and the UAV will transmit its data [6, 7].

6.4.2 Changing Topology

In order to cope with the frequent connection loss between nodes in a FANET, as mentioned earlier, two types of routing protocols have been selected.

6.4.2.1 Cluster based

One routing technique that has been proposed in two variants is the use of clusters. This routing protocol assumes a significant number of nodes. The first variant does not have a hierarchy. All the nodes are divided into clusters and each cluster has a relay member. All nodes maintain a list of their neighbours, a list of clusters in the network and a list of boundary nodes. The relay member will send the data to the designated clusters. The authors of [12] emphasize that the chosen route may not be the shortest one. The second variant makes use of a hierarchy. This is Multi-Level Hierarchical Routing. A FANET is divided into clusters. Each cluster has a cluster head (CH). The CH is connected to the upper- and lower layers of the network. If UAVs wish to send data to another UAV they send it to the CH, which will relay it to the designated UAV in the other layer. The CH needs to be in range of all the UAVs that are part of its cluster [12, 13].

6.4.2.2 Dynamic source protocol

The Dynamic Source Routing Protocol (DSR) is a reactive protocol. This means that UAV nodes update their routing tables when they need to communicate with a certain node. The intention of the DSR is to reduce the overhead bandwidth in the network. Furthermore it should increase battery power, because the decrease in routing messages reduces the workload of the CPU. When a node wishes to transmit to a certain node it will use the route discovery protocol. The transmitting node will send out a route request with an ID and an empty list. On this list all the intermediate nodes that lie on the route to the target node will be added. Intermediate nodes that receive this request will, if they are connected to the target node, send it to the target node. If the intermediate nodes are not connected to the target node, they will send it to other non-target nodes in their vicinity. The intermediate node will also append its own name to the intermediate-node list that is part of the route request. If an intermediate node receives another route request that has the same ID or the nodes name is already on the list, the intermediate node will discard the request. When the target node receives the route request it will send a reply back to the transmitting node. The transmitting node now knows which route its messages have to take and will start transmitting its data. The route the node has now discovered, will be saved on the nodes cache memory. Furthermore, while sending the data the transmitting node will receive messages from intermediate nodes on whether they have received the data. This is called route maintenance and it should inform a transmitting node about whether a route still exists [4, 13–15].

6.4.3 Clock Synchronization

Since it has become apparent that timing-beacons may not get through, due to contention, the Adaptive Timing Synchronization Procedure (ATSP) has been proposed in [16].[1] It has been observed that only faster timing synchronizes others. Therefore, ATSP gives priority to the timing-beacons from the fastest nodes. The slower nodes will have the frequency of their timing beacons reduced. Though ATSP is able to synchronize nodes timing differences of hundreds of microseconds remain. In addition, the performance of ATSP decreases when the number of hops increases. In [17] the authors propose the Automatic Self-time-correcting Procedure (ASP).

---

[1]Note that ATSP has also been proposed for the IEEE 802.11/wifi protocol.

This procedure incorporates the procedure of ATSP, but after the slower nodes have synchronized the priority of their timing beacons is increased. This allows the timing information to spread throughout the rest of the network, enabling other nodes, not in the vicinity of the faster nodes, to synchronize.

6.4.4 Security

One method to authenticate UAVs in a FANET is to use a network backbone made up of UAVs that would provide authentication services for the network. This method could, however, lose its authentication service if those UAVs are destroyed or malfunction. In [11] adaptive security is proposed as a way to address possible loss of the backbone. If the backbone is rendered inoperable, the nodes switch to an infrastructure-less mode in which the authentication services are distributed among the nodes. When the nodes initiate a connection with each other, they check each other's certificates on whether they have expired and whether they are on a revocation list. The revocation list includes all certificates that cannot be trusted. If a node uses a revoked certificate, other nodes will not communicate with that node. When the backbone is operational again, authentication is again done via the backbone.

6.5 Discussion

6.5.1 Aircraft Altitude and Mobility

The speed with which UAVs manoeuvre and the rapid altitude changes they can make, pose an issue for developers as to whether they are to use omnidirectional or directional antennas in the designs of their FANETs. The solution that has been presented is to use a combination of antennas with a mac-protocol that has been adapted to it. The solution could work, but there will be particularly limitations for the military operational use of a FANET with such design. In [7] the authors themselves acknowledge that in order to use the directional antenna, the UAV that will receive the data needs to move to a predefined location. While communicating the receiving UAV will also need to stay within a certain angle of the transmitting UAV. Otherwise it would move out of the beam of the transmitting antenna. These requirements will restrict the mobility of the UAVs, making them unsuitable, for example, in flying at high speeds.

6.5.2 Changing Topology

The rapid manoeuvres, high speed and possibility of losing UAVs due to malfunctions or enemy fire will result in frequent loss of connection between UAVs, thereby changing the topology of the FANET. The cluster routing protocol is proactive in updating its routing tables and the dynamic resource protocol is reactive. The most appropriate routing protocol probably depends on the size of the FANET and the available bandwidth. The cluster routing protocol is more appropriate for large FANETs, because UAVs only have to communicate via a CH and receive their routing table updates via the CH. This leads to a reduction in overhead, which is necessary for FANETS of significant size. The CHs, however, have to be in range of all the UAVs within its cluster and the loss of the CH would lead to a partition of the FANET. The DSR is more suitable for smaller FANETs. Each UAV will only look for an appropriate route when it needs to send data. In a small FANET this could reduce the overhead. In a larger FANET it would take a longer time for UAVs to find a route to their target UAV. Moreover, UAVs would have to respond to the route request and this increase in broadcasts could lead to more contention.

6.5.3 Clock Synchronization

Large FANETS that use frequency hopping could have trouble with their clock synchronization, due to contention problems. ATSP, which has faster nodes given higher priority to their timing beacons over those of slower nodes, was presented. ASP incorporated ATSP but also reverses the priority to enable the rest of the network to synchronize. Research in [17] has shown that the technique works, even with large networks, and could therefore solve the issue of clock synchronization [17].

6.5.4 Security

It was concluded earlier that the use of symmetric keys for authentication in a large FANET would not work. The use of asymmetric keys would require an infrastructure that would be prone to single-point-of-failure. The authors of [11] proposed a solution by introducing an infrastructure-less mode in which the nodes would carry out the authentication themselves. The simulations carried out show a high rate of successful requests and a limited delay in setting up a connection with both sides authenticated.

6.6 Conclusion

In conclusion, the issues of clock synchronization and security seem to be possible to solve using the technologies presented in this chapter. The solution to the high mobility of UAVs and differences in altitude is not satisfying. Although the use of directional antennas to send data is less susceptible to interference and jamming, it also requires UAVs to move to designated spots to make communication possible. This limits the operational use of FANETs that use directional antennas. The last issue, changing topology, appears to require solutions based on the type of FANET used. If the FANET is large, it could use a cluster based protocol that has all the data from nodes send to the cluster head, which will relay the data to other clusters. If the FANET is small, it could use the Dynamic Source Routing protocol, which has all nodes find their route to their target node only when they need to communicate, resulting in less overhead. All in all the problems facing the development of FANETs seem difficult, due to the characteristics of FANETs, but the technologies seem to be able to do the job.

References

[1] J. Lin and P. W. Singer, "China is making 1,000-UAV drone swarms now," Popular Science, 9 January 2018. [Online]. Available: https://www.popsci.com/china-drone-swarms. [Accessed 19 January 2018].

[2] J. L. Miere, "Russia Developing Autonomous 'Swarm of Drones' In New Arms Race With U.S., China," Newsweek, 15 May 2017. [Online]. Available: http://www.newsweek.com/drones-swarm-autonomous-russia-robots-609399. [Accessed 19 January 2018].

[3] D. Hambling, "Drone swarms will change the face of modern warfare," Wired, 7 January 2016. [Online]. Available: http://www.wired.co.uk/article/drone-swarms-change-warfare. [Accessed 19 January 2018].

[4] T. Brown, B. Argrow, C. Dixon, S. Doshi, R. G. Thekkekunnel and D. Henkel, "Ad Hoc UAV Ground Network (AUGNet)," in *Unmanned Unlimited Technical Conference, Workshop and Exhibit*, Chicago, 2004.

[5] I. Bekmezci, O. K. Sahingoz and S. Temel, "Flying Ad-Hoc Networks (FANETS)," 2013. [Online]. Available: http://dx.doi.org/10.1016/j.adhoc.2012.12.004. [Accessed 17 January 2018].

[6] A. I. N. Alshbatat, "Cross-Layer Design For Mobile Ad-Hoc Unmanned Aerial Vehicle Communication Networks," Dissertations, Western Michigan University, 2011.

[7] A. I. Alshbatat and L. Dong, "Adaptive MAC Protocol for UAV Communication Networks Using Directional Antennas," in *Networking, Sensing and Control (ICNSC)*, Chicago, 2010.

[8] L. Gupta, R. Jain and G. Vaszkun, "Survey of Important Issues in UAV Communication Networks," *IEEE Communications Surveys and Tutorials,* vol. PP, no. 99, 2015.

[9] C. M. Chao, J. P. Sheu and I. C. Chou, "An Adaptive Quorum-Based Energy Convering Protocol for IEEE 802.11 Ad Hoc Networks," *IEEE Trans. on Mobile Computing,* vol. 5, no. 5, pp. 560–570, 2006.

[10] J. Mick, "Iran: Yes, We Hacked the US's Drone, and Here's How We Did It," Daily Tech, 15 December 2011. [Online]. Available: http://www.daily tech.com/Iran+Yes+We+Hacked+the+USs+Drone+and+Heres+How+We +Did+It/article23533.htm. [Accessed 17 January 2018].

[11] J. Kong, H. Luo, K. Xu, D. L. Gu, M. Gerla and S. Lu, "Adaptive security for multilevel ad hoc networks," *Wireless Communications and Mobile Computing,* no. 2, pp. 533–547, 2002.

[12] P. Krishna, N. H. Vaidya, M. Chatterjee and D. K. Pradhan, "A cluster based approach for routing in dynamic networks," *ACM SIGCOMM Computer Comunication Review,* vol. 27, no. 2, pp. 49–64, 1997.

[13] H. H. Okulu, "Networking Models in Flying Ad-Hoc Networks (FANETs): Concepts and Challenges," *Journal of Intelligent and Robotic Systems,* no. 4, 2014.

[14] D. B. Johnson and D. A. Maltz, "Dynamic Source Routing in Ad Hoc Wireless Networks," in *Mobile Computing*, Kluwer Academic Publishers, 1996.

[15] D. B. Johnson, D. A. Maltz and J. Broch, "DSR: the dynamic source routing protocol for multihop wireless ad hoc networks," in *Ad hoc networking*, Boston, Addison-Wesley Longman Publishing Co. Inc., 2001, pp. 139–172.

[16] L. Huangand and T. H. Lai, "On the Scalability of IEEE 802.11 Ad Hoc Networks," in *Third ACM International Symposium on Mobile Ad Hoc Networking and Computing*, Lausanne, 2002.

[17] J. P. Sheu, C. M. Chao and C. W. Sun, "A Clock Synchronization Algorithm for Multihop Wireless Ad Hoc Networks," *Wireless Personal Communications,* no. 43, pp. 185–200, 2007.

7

Moving Object Detection: Deep Learning Approach with Security Challenges

Shivprasad P. Patil[1] and Ramjee Prasad[2]

[1] NBN Sinhgad School of Engineering, Pune, India
[2] CTIF Global Capsule (CGC), Department of Business Development
and Technology, Aarhus University, Herning, Denmark
E-mail: shivprasad.patil@sinhgad.edu; ramjee@btech.au.dk

In many computer vision applications, robust and real-time foreground detection is a crucial issue. There has been considerable work on object detection and activity monitoring. Wherein Background subtraction is typically used approach for foreground detection. It is found that, this approach is normally useful for stationary camera but does not work well for rotating camera. The challenge in this case is to segregate the dynamic entities while the background rotates virtually at a fixed pace.

Further, even in capturing outdoor scenes by a fixed camera, the camera cannot completely be considered as stationary due to the uncontrolled environment, such as jitter problems in camera or camera shake problems. Thus, the increasing use of moving cameras along with growing interests in detecting moving objects make it essential to develop robust methods of moving object detection for moving or rotating cameras. The benefits that we derive from this method are more accurate detection of the actual moving object, thereby eliminating nonessential data, which is redundant.

This chapter presents some conventional and recent technical directions towards overcoming the challenges in developing the robust and secure methods of moving object detection.

Security within CONASENSE Paragon, 95–104.

7.1 Introduction

In many computer vision (CV) applications, accurate and real-time foreground (FG) detection is a critical issue. Accuracy of FG algorithm affects the accuracy and computation performance of later stages. It is known that outdoor environment is more complex than indoor one. Hence, FG detection is easier in indoor environment than outdoor one. Ideally we expect FG algorithms to work well in both cases. There has been considerable work on these issues [1]. However, such work assumes non-rotating camera. In contrast, we address the problem of moving object detection over a wide area through a rotating sensor. Motivation for use of this type of sensor is ease of installation and maintenance, thereby saving of cost. Also, processing and transmission "overheads" in visual surveillance applications are reduced.

Motion detection with a non-stationary viewing sensor has received considerable attention of the researchers. In many CV applications robust and accurate foreground segmentation is a crucial issue [1]. There are numerous applications which require FG segmentation. In them, most prominent one is automated visual surveillance [2–4]. It is found that, existing background subtraction (BS) algorithms cannot be used directly for the sequence coming from rotating camera. To compensate for the motion due to the moving sensor, motion compensation is needed. Usually a motion model of the background is assumed, and motion parameters are estimated. Then the background is registered ideally and the foreground can be detected on pixel level.

The use of background modelling for detecting the moving object is very common in many applications. In the scene like video surveillance, it is difficult to achieve the background model that can be developed by acquiring a background image which doesn't include stationary object. In some situations there is a change in illumination conditions. Also, object is removed or introduced from the scene. Many background modelling methods have been developed, by considering the problems cited above to make them more robust and adaptive. Although most of these methods use only a fixed camera, they provide a good starting point for a rotating camera.

Many algorithms have been proposed for object detection in video surveillance applications. They rely on different assumptions e.g., statistical models of the background [5, 6], minimization of Gaussian differences [7], minimum and maximum values [8], adaptivity [9] or a combination of frame differences and statistical background models [10].

The authors of this chapter have presented a novel approach of video coding technique for accurate estimation of object detection in video surveillance [11]. In this approach, a least mean estimator for denoising is used and recurrent full search motion estimator logic is defined for the prediction of foreground moving elements from the video sequence, which comes from a rotating sensor.

Recently, due to the availability of annotated video datasets, deep Convolutional neural networks (CNNs), which require exhaustive training, have been considered for moving object detection for moving cameras [12]. In this approach, a CNN is learned from data to select meaningful features and estimates an appropriate background model from video. Their approach is capable to be used in real time applications for moving object detection under various video scenes.

Summary of prominent challenges in moving object detection

- It is very challenging to evaluate the background model when sensory camera is moving or rotating.
- During the evaluation of background frame, the process mainly gets affected due to occlusion and illumination changes.
- Difficulty in segmentation process due to the presence of background clutter. Thus, it is impossible to represent a background and divide the moving foreground objects.
- Shadows transmitted from foreground objects leads to the difficulties in processing with respect to background subtraction. Hence, the overlapping shadows delays their partition and classification.
- Background subtraction techniques have to deal with the signal that gets corrupted by various noises such as sensor noise and compression artifacts.

7.2 Background Subtraction

The majority of background subtraction algorithms are comprises of several processing modules, as shown in Figure 7.1, which are explained subsequently.

Background Model: The background model is an important part of the background subtraction algorithm. Normally, the background model is used as a reference to compare with the incoming video frames. The initialization of the background model plays important role since video sequences are normally not completely free of foreground objects during the BG model development phase. As a consequence, the model gets corrupted by including

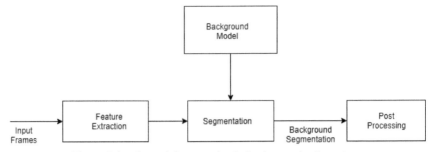

Figure 7.1   General framework of a background subtraction.

foreground objects into the background model, which leads towards false identification.

Feature extraction: In order to compare the background frame with the incoming video frames, adequate features that represent relevant information must be chosen. Most algorithms use RGB intensities or gray scale as features. In some cases, pixel intensities along with other features are combined. Also, the choice of the feature region is vital. One can extract the features over pixels, blocks or patterns. Pixel-wise features often yield noisy segmentation results since they do not encode local correlation, while block-wise or pattern-wise features tend to be insensitive to slight changes.

Segmentation: With the help of a background model, the incoming video frames can be processed. Background segmentation is performed by extracting the features from corresponding pixels or pixel regions of both frames and using a distance measure, e.g. the Euclidean distance, to measure the similarity between those pixels. After being compared with the similarity threshold, each pixel is either labeled as background or foreground.

Post-Processing: The spatial-median filtering, which is a commonly used post-processing method in background subtraction, returns the median over a neighborhood of given size (the kernel size) for each pixel in an image. As a consequence, the operation removes the outliers in the segmentation map and also blob smoothing is achieved.

7.3 Deep Learning (DL)

Deep learning has become popular since 2006, due to a major breakthrough in speech recognition [13]. It outperformed Hidden Markov Model and Gaussian Mixture Model (HMM-GMM) with a huge margin, which dominated the field for many years. There are few reasons making neural networks successful again. First of all, a key reason is the emergence of large scale training data with annotations. Moreover, there have been significant advances in the design of network structures, models, and training strategies. With large-scale training data, deep neural networks show significant advantages compared with shallow models because of their very large learning capacity.

7.3.1 Background

Success story of deep learning lies in the advancements of deep neural networks (DNNs) and fast development in high performance parallel computing systems, trained with huge amount of data (big data). It is imperative to know the components and the recent DL models that are widely used. Various DNN model architectures are shown in Figure 7.2.

Convolutional Neural Network (CNN): A CNN consists of one or more convolutional layers, which use convolutional operations to compute layer-wise results. This operation enables the network to learn about spatial information through training and hence CNNs show outstanding performances especially on vision applications.

Recurrent Neural Network (RNN): A recurrent neural network (RNN) is widely used to process sequential data. As illustrated in Figure 7.2(b), an RNN updates the current hidden unit and calculates the output based on the current input and past hidden unit.

Generative Adversarial Network (GAN): A generative adversarial network (GAN) framework consists of a discriminator D and a generator G. G generates fake data while D determines whether the generated data is real. Usually generators and discriminators are neural networks that can have various structures depending on the application. GANs are actively studied in various ftelds, such as image/speech synthesis.

7.3.2 DL and Object Detection

More recently, a novel approach for background subtraction with the use of convolutional neural network (CNN) is proposed [15]. This approach uses a

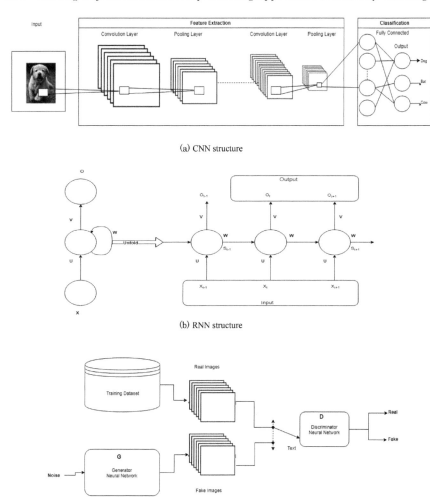

(a) CNN structure

(b) RNN structure

(c) GAN Structure

Figure 7.2   Deep Neural Network model structures [14].

fixed background model, which is generated from a temporal median operation over some video frames. Afterwards, a scene-specific CNN is trained
with corresponding image patches from the background image, video frames
and ground truth pixels, or alternatively, segmentation results coming from
other background subtraction algorithms. After extracting a patch around a
pixel, feeding the patch through the network and comparing it with the score
threshold, the pixel is assigned with either a background or a foreground

Figure 7.3 Framework of a background subtraction using deep learning.

label. After reassembling the patches into the complete output frame, it is post-processed, which yields the final segmentation of the respective video frame. This technique is illustrated in Figure 7.3.

However, due to use of highly redundant data for training, the network is scene-specific, i.e. can only process a certain scenery, and needs to be retrained for other video scenes (with scene-specific data).

7.4 Security Aspects of Deep Learning

Deep learning and neural networks have become very prominent in shaping the technology that powers various industries. From content recommendation to disease diagnosis and treatment and self-driving vehicles, deep learning is playing a very vital role in making important decisions. Therefore, it is essential to know the unique security threats of deep learning algorithms

Deep learning algorithms have two characteristics that are relevant from a cybersecurity perspective:

- They are overly reliant on data, which means they are as good (or bad) as the data they are trained with.
- They are opaque, which means we don't know how they function (or fail).

It is interesting to see how malicious actors can use the unique characteristics of deep learning algorithms to stage cyberattacks.

7.4.1 Adversarial Attacks

Malicious actors can leverage these mistakes to stage adversarial attacks against systems that rely on deep learning algorithms. For instance, by making small tweaks to stop signboard, it cannot be sensed or detected by computer vision algorithm that is used by self-driving cars. This means that a hacker can force a self-driving car to behave in dangerous ways and possibly cause an accident. Adversarial attacks are not limited to computer vision. They can also be applied to voice recognition systems or any other system that rely on deep learning.

7.4.2 Data Poisoning

While adversarial attacks find and abuse problems in deep learning, data poisoning creates problematic behavior in deep learning algorithms by exploiting their over-reliance on data. Deep learning algorithms have no notion of moral, commonsense and the discrimination that the human mind has. They only reflect the hidden biases and tendencies of the data they train on. Because deep learning algorithms are only as good as their data, a malicious actor that feeds a neural network with carefully tailored training data can causing it to manifest harmful behaviour. This kind of data poisoning attack is especially effective against deep learning algorithms that draw their training from data that is either publicly available or generated by outside actors.

There are already several examples of how automated systems in criminal justice, facial recognition and recruitment have made mistakes because of biases or shortcomings in their training data. While most of these examples are unintentional mistakes that already exist in our public data due to other problems that plague our societies. There's nothing preventing malicious actors from intentionally poisoning the data that trains a neural network.

Based on the knowledge of the structure and parameters of the deep learning model, attacks on DL models usually attempt to subvert the learning process or induce false predictions on the purpose, by injecting adversarial samples. This type of attack, which can include gradient-based techniques, is often called a white-box attack. In contrast, black-box attacks lead the target system to make false predictions, without any information about the underlying model. Black box attack is a replicating a target model without knowledge of the architecture of the target model. It is observed that most of the attacks exploit the prediction confidence given by the targeted model without knowing the model's structure and parameters.

7.4.3 Data Privacy

DL requires users to transfer some sensitive data to remote machines because of the computational cost or the need for collaborative training. In such situations, users lose control over the data after the transfer and have concerns about their data privacy being stolen between transfers, or the service holders that they upload their data to, can misuse their data without consent. Therefore, it is essential to have some privacy-preserving techniques in place. Most recent privacy preserving techniques are homomorphic encryption and differential privacy.

7.5 Summary

In order to extract moving object from video sequences in real time, this chapter presented several technical directions towards accurate and efficient detection of moving object. Initially, this chapter described conventional methods of object detection, along with its constraint/limitations. Further, very recent trend of deep learning for object detection is described. Security and privacy challenges are also discussed. Authors believe that, this chapter will help in the future research activity, especially when video sequence is generated from rotating camera.

References

[1] J. Cheng, J. Yang, Y. Zhou and Y. Cui., " Flexible background mixture models for foreground segmentation" in Proceedings of Image and Vision computing, 24, 473–482, 2006.

[2] Christof Ridder, OlafMunkelt, and harald kirchner "Adaptive background estimation and foreground detection using Kalman-Filtering" in Proceedings of the International Conference on recent Advances in Mechatronics, ICRAM '95, UNESCO Chair on Mechatronics, 193–199, 1995.

[3] Stauffer C, Grimson W., " Adaptive background mixture models for real-time tracking" in Proc IEEE Conf on Comp Vision and Patt Recog (CVPR 1999), 246–252, 1999.

[4] Cheung S, Kamath C, " Robust background subtraction with foreground validation for Urban Traffic Video" in Journal Appl Signal Proc, Special Issue on Advances in Intelligent Vision Systems: Methods and Applications (EURASIP 2005), New York, USA, 2005; 14:2330–2340. M. Young, The Technical Writer's Handbook. Mill Valley, CA: University Science, 1989.

[5] C. Stauffer, W. Eric, and L. Grimson, "Learning patterns of activity using real-time tracking," IEEE Trans. Pattern Anal. MachineIntell., vol. 22, no. 8, pp. 747–757, August 2000.

[6] T. Bouwmans, F. Baf and B. Vachon. "Background modeling using mixture of gaussians for foreground detection – A survey." in Recent Patents on Computer Science 1, 3, pp. 219–237, 2008.

[7] N. Ohta, "A statistical approach to background suppression for surveillance systems," in Proceedings of IEEE Int. Conference on Computer Vision, pp. 481–486, 2001.

[8] I. Haritaoglu, D. Harwood, and L. S. Davis, "W4: Who? when? where? what? a real time system for detecting and tracking people," in IEEE International Conference on Automatic Face and Gesture Recognition, pp. 222–227, April 1998.

[9] M. Seki, H. Fujiwara, and K. Sumi, "A robust background subtraction method for changing background," in Proceedings of IEEE Workshop on Applications of Computer Vision, pp. 207–213, 2000.

[10] R. Collins, A. Lipton, and T. Kanade, "A system for video surveillance and monitoring," in Proc. American Nuclear Society (ANS) Eighth Int. Topical Meeting on Robotic and Remote Systems, Pittsburgh, PA, pp. 25–29, April 1999.

[11] S. Patil, R. Sanyal and R. Prasad, " Efficient video coding in region prediction in online video surveillance" in Int'l Conf. IP, Comp. Vision, and Pattern Recognition (IPCV'15), pp. 210–216, 2015.

[12] M. Babaee, D. T. Dinh, and G. Rigoll, "A deep convolutional neural network for background subtraction" in arXiv preprint arXiv:1702.01731, 2017.

[13] X. Wang., "Deep Learning in Object Recognition, Detection, and Segmentation" in Foundations and Trends R in Signal Processing, vol. 8, no. 4, pp. 217–382, 2014.

[14] H. Bae, J. Jang, D. Jung, H. Jang, H. Ha, and S. Yoon, "Security and Privacy Issues in Deep Learning" in ArXiv :1807.11655, July 2018.

[15] M. Braham and M. Van Droogenbroeck, "Deep background subtraction with scene-specific convolutional neural networks" in *International Conference on Systems, Signals and Image Processing, Bratislava* 23–25 May 2016.

8

The Network Neutrality in Service Innovation Era

Yapeng Wang[1] and Ramjee Prasad[2]

[1]International Cooperation Department, China Academy of Information and Communications Technology, MIIT, China
[2]Center for TeleInFrastruktur (CTIF), Arhus University, Denmark
E-mail: wangyapeng@caict.ac.cn; ramjee@btech.au.dk

In 2015, Federal Communication Committee (FCC) and European Commission enacted respective rules to open internet, the network neutrality (as commonly known) or Net Neutrality (as in Europe) was gained more attentions and the topic was discussed ardently. On December 14, 2017, the FCC voted 3: 2 to repeal "Internet Open Regulations." This chapter reviews the NN debate process, and the opinions from relevant sides, including the network providers, the service providers, other relevant companies, governments and researchers. This chapter also introduces a telecommunication convergence concept that is CONASENSE (Communication, Navigation, Sensing and Services). It aims to formulate a vision on solving societal problems with new telecom technique to improve human welfare benefit. This chapter focuses on the service of CONASENSE and made some analysis the NN in service innovation era.

8.1 Introduction

The Information and Communication Technology (ICT) is playing an absolutely important role in our life and with more and more innovative internet services and applications emerge, from E commerce, e-health to a real-time telephone meeting and a live video streaming, which are improving human's Quality of Life and benefiting the whole society. We have entered a service innovation era, in which every part of the industry chain is making contribution to the telecom industry. Internet Services Providers (ISPs) upgrade the network, the Content Providers (CPs) keep the telecom industry prosperous. How these innovative service run over the telecommunication network is governed by not only technology, but also by the rules as proved by government or authorities in some region. NN, as a regulating rule, aims that every end user have the equal right to access the internet and use the legal internet content and applications. On December 14, 2017, the Federal Communications Commission (FCC) abolished the neutral network policy and re-empowered telecom operators with control over broadband Internet access. This neutral network policy was repealed and it can also be regarded as the official opening of the Light-touch Regulation era in the United States. CONASENSE, as will be discussed later in this chapter, is a telecom convergence concept that will be run above ICT platform. This chapter will focus on the services in CONASENSE, and analyze the abolishment of NN rules impact on those services.

8.1.1 NN

Network Neutrality rules aims to provide an open internet [2] to the end users. The open internet is defined by FCC and refers to *"uninhibited access to legal online content without broadband Internet access providers being allowed to block, impair, or establish fast/slow lanes to lawful content"* [1].

This means that a legitimate content, whether it is an application or data, must reach users without the intermediate communication system controlling its flow. Internet Content Providers (ICPs) and Internet Service Providers (ISPs) cannot block, throttle or create the special facilities for a content or application. This gives a kind of liberty to the end users to enjoy the variety of information without bothering about how ICT is dealing with that information. Therefore, users may demand a better network services to enjoy the lawful contents. Section 8.2 will introduce NN in detail.

8.1.2 Innovative Services

In this chapter, the innovative service refers to over the top service (OTT) which implies communications carried over the physical network infrastructure using an IP protocol to reach services available on the internet [3]. An OTT application is any app or service that provides a product over the Internet and bypasses traditional distribution. Services that come over the top are most typically related to media and communication and are generally, if not always, cheaper than the traditional method of delivery. Section 8.4 will discuss how the change of NN impact on it.

8.1.3 CONASENSE [4]

CONASENSE foundation was established as a brain tank in November 2012. CONASENSE refers to the Communication, Navigation, Sensing and Services. Its main aim is to define and steer processes directed towards actions on investigations, developments and demonstrations of novel CONASENSE services, as an integration of communications, navigation and sensing technology, with high potential and importance for society. CONASENSE has a large research scope and it has a huge capacity to develop various application and service provision to users, as shown in Figure 8.1. Section 8.4 will give a detail introduction to this concept.

This chapter points out that both sides (both sides refer to the opponent and proponent of NN in this chapter) of the NN debate agree on the need to preserve the Internet as a space that is open to innovation, and the freedom of users to access the content and services. Meanwhile this chapter also analyses how NN impact innovation such as CONASENSE, and meet the regulation goal. This chapter makes analysis mainly from technique perspective, and the social and economic aspects are also discussed.

Apart from Section 8.1, the rest of the chapter is organized as follows: Section 8.2 reviews the evolution and current discussions of NN,

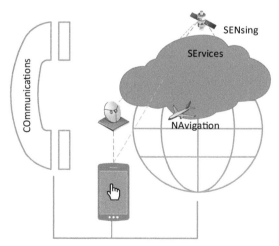

Figure 8.1 CONASENSE Framework [4].

Figure 8.2 The Countries that either have enacted or are in their discussions.

Section 8.3 briefly introduces CONASENSE, Section 8.4 qualitatively analyses the impact of the abolishment of NN on Services and finally Section 8.5 concludes the chapter.

8.2 The Network Neutrality

In the world, more and more regulators believe that the CPs are the motivation of innovation, economy and investment, but with the rapid development of the Internet as an ubiquitously available platform and resource, the network infrastructure owners are regarded as to the broadband providers are regarded as to "*have both the incentive and the ability to act as gatekeepers standing between edge providers and consumers. As gatekeepers, they can block access; target competitors, extract unfair tolls*" [1], but the open internet is regarded to the guarantee to the innovation, economy and investment. This is one of the vital reasons why more and more governments enact the NN rules. Presently, more than 10 countries enacted relevant rules [5]. Figure 8.2 shows the countries that either have enacted NN or are in their discussion phase. On December 14, 2017, the FCC Committee voted 3: 2 to repeal "Internet Open Regulations."

8.2.1 The Concept NN

NN commonly indicates that Internet services providers make or keep the Internet open and ensure all the users have same right to access to the network and use the content and services without any discrimination [1].

8.2.2 The Principles of NN

Currently, Network Neutrality is a global debate [5]. FCC released its updated Open Internet Order in 2015, to enact strong, sustainable rules to protect the Open Internet. The order includes 3 bright-line rules as [1]:

- No Blocking (NB), to prohibit the network providers to block the legal applications, contents and devises.
- No Throttling (NT), to prohibit the network providers to degrade the traffic of legal applications and contents.
- No Paid Prioritization (NPP), to prohibit the network providers to provide and charge the differentiated service for the applications and contents.

"No unreasonable interference or disadvantage to consumers or edge providers" [1], and, enhanced transparency between users and ISPs. "A*s with the 2010 rules, this Order contains an exception for reasonable network management, which applies to all but the paid prioritization rule*" [1] is also emphasized in FCC rules.

Nonetheless, in October 2015, the European Parliament approved the first EU-Wide rules on NN that enshrined its principle into EU law [6] that says: "*No blocking or throttling of online content, applications and services. Accordingly, every European must be able to have an access to the internet and, all contents and services, via a high-quality service that is provided by an ISP such that all traffic must be treated equally. NN rules seems more inclined to the end users and, equal treatment allows reasonable day-to-day traffic management according to justified technical requirements which must be independent of, the origin or destination of the traffic, and, of any commercial considerations.*"

8.2.3 The State of the Art of NN

On December 14, 2017, the Federal Communications Commission (FCC) abolished Obama's neutral network policy and re-empowered telecom operators with control over broadband Internet access. The FCC said, the former network neutral "three ban", mandatory broadband access behavior, no longer supervise the future. FCC requires relevant sides strengthen self-regulation, and regulators will increase their own transparency.

The decree aroused the concern of all parties, the U.S. telecommunications industry applauded the neutrality repeal order, but the Internet industry expressed dissatisfaction. In fact, the abolition of the network neutrality policy is only a silhouette of the FCC to adjust the regulatory thinking of the telecommunications industry, accompanied by changes in the direction of a series of core policies, such as the classification of telecommunications services, the FCC on its own power reform and many other issues. In fact, since the presidential election in early 2017 and the Republican Party came to power, the telecommunications industry has been making good progress. Neutral network policy was repealed and it can also be regarded as the official opening of a new Light-touch Regulation era in the United States.

8.2.4 History of NN

The phrase of NN was first proposed in a law review article [7] by Tim WU in 2003, introduced the concepts of freedom, competition and innovation. NN was only a concept suggesting that each network protocol layer should be independent and perform the assigned duties at the original phase. With the expansion of the commercial-scale internet, the focus of the competition has been shifting from the connection and network layer to the application and content layer which made the debate of NN switch from technological field to commercial field. Because the term holds different meanings to people in different field, it has different debate focuses in different process phases. Although NN is now abolished, the debating still continues.

FCC Open Internet Order raised several rounds of debates. Through the key event happened in U.S as follows:

- In December 2010, The FCC approved the Open Internet Order which was consisted of three items of NN regulations, and they are *"Transparency, No Blocking and No unreasonable discrimination"* [8].
- In January 2014, United States Court of Appeals for the District of Columbia Circuit (D.C. Circuit) overturned the Open Internet Order.
- In February 2015, The FCC issued Open Internet Rules and Order. And In June 2015, The Open Internet Rules and Order came into effect officially.
- In August 2015, The D.C Circuit announced the Open Internet Rules and Order will face an important federal-court test;
- On December 14, 2017, the Federal Communications Commission (FCC) abolished Obama's neutral network policy and re-empowered telecom operators with control over broadband Internet access.

8.2.5 The Analysis of Abolition of NN

On December 14, 2017, the FCC Committee voted 3: 2 to repeal "Internet Open Regulations." The main contents include: First, abolish network neutrality policy, relax regulation, restore freedom of broadband access market. The FCC states that a network-neutral policy of "heavy regulation" of broadband Internet access services (BIAS) will increase the potential operating costs of the entire Internet ecosystem. Therefore, the FCC will regain the "light regulatory" approach 20 years ago to stimulate growth and restore market liberalization and freedom. Followed by the service classification adjustment. The BIAS service is re-classified under the Chapter "Information Services" under "Telecommunications Business" of the "Communications Act of 1934".

This adjustment will bring about four changes. First, the FCC no longer regulates the BIAS service, including both mobile and fixed broadband access. Secondly, the broadband access services launched by some Internet giants in the United States, such as the Google Fiber business conducted by Google, were previously included in the regulation of "telecom services" (and even triggered a "whether Google Inc. is an Internet company or a telecommunications company Enterprise" argument). After the current round of adjustments, the return of such businesses carried out by Internet companies falls into the category of "information services" in Chapter 1 of the "Communications Act of 1934". Third, the power limit and part of the power out. The FCC banned state regulators from legislating for the network neutrality or issuing policies. In addition, the power of broadband consumer protection, Internet data security protection and other powers transferred to the Federal Trade Commission (FTC). Finally, ISPs are required to be self-disciplined, increase the transparency of information disclosure and require disclosure to consumers and government regulators of how they conduct paid-for-network access services.

According to the "Executive Decree on Rescheduling Internet-Related Policies at the End of November 2017" (WC Docket No. 17-108), the FCC has released the main reasons for abrogating network neutrality. First of all, the basic attributes of the network access service are changed. FCC believes that the broadband access service provided by telecom operators should be a valuable commodity rather than a public product and should not be regarded as a public service such as water, electricity and gas, and should not be used for public service undertakings "Heavy regulation" approach. Therefore, in theory, telecom operators have the right to arrange their own business operations, the government should not supervise. Second, the enthusiasm of telecom companies was hit and broadband investment declined. Since the introduction of the network-neutral policy by the FCC in February 2015, U.S. investment in broadband infrastructure has dropped for the first time since 2009, while the total investment in broadband of AT & T, Verizon, Comcast fell 5.6%. For such results, the United States industry, economics and FCC believe that the network neutral policy is the direct reason to combat the investment enthusiasm of telecom operators.

8.2.6 Current Discussion of NN

From the history of Network Neutrality, we can find that NN has been really controversial. Although US abolish NN, The essential debate in all over the world will continue and the focus point will be discussed below:

8.2.6.1 Service innovation

The proponents of NN are mainly those enterprises that are related to Internet contents. Worriedly, they are stating that actions departing from NN principles could threaten the innovation of the internet content as, service providers may increase control on the content and applications over the internet. In order to encourage the innovative services with enhance quality of service especially from startups, the new EU net neutrality rules *"enable the provision of specialized or innovative services on condition that they do not harm the open internet access. These services use the internet protocol and the same access network but require a significant improvement in quality or the possibility to guarantee some technical requirements to their end-users that cannot be ensured in the best effort open internet* [9]. *These specialized or innovative services have to be optimized for specific content, applications or services, and the optimization must be objectively necessary to meet service requirements for specific levels of quality that are not assured by the internet access services."* The rules also urges these services cannot be a substitute to internet access service, can only be provided if there is sufficient network capacity and must not be to the detriment of the availability or general quality of internet access service for end-users.

FCC's rules refer above mentioned service as non-Broadband Internet Access Service [1] *"(non-BIAS) data services, which are not subject to the rules. According to the rules, non BIAS data services are not used to reach large parts of the Internet, not a generic platform–but rather a specific "application level" service, and use some form of network management to isolate the capacity used by these services from that used by broadband Internet access services."*

8.2.6.2 The investment on the network infrastructure

The opponent of NN which is typically network operators argue that NN regulation will make it more difficult for Internet service providers (ISPs) and other network operators to recoup their investments in broadband networks, weaken the incentives to invest and upgrade the telecom infrastructures. Some network provider has argued that they will have no incentive to make large investments to develop advanced fibre-optic networks if they are prohibited from charging higher preferred access fees to companies that wish to take advantage of the expanded capabilities of such networks [10]. FCC reclassified the BIAS as telecommunication service in the Open internet Order [1], FCC believed that the reclassification will preserve investment incentives.

8.2.6.3 The management of internet traffic by internet service providers and what constitutes reasonable traffic management

Commonly, "traffic management is used to effectively protect the security and integrity of networks. It helps deal with temporary or exceptional congestion or to give effect to a legislative provision or court order. It is also essential for the certain time-sensitive service such as voice communications or video conferencing that may require prioritization of traffic for better quality. But there is a fragile balance between ensuring the openness of the Internet and the reasonable and responsible use of traffic management by ISPs [11]. The opponent of NN argued that NN may prove ineffective in such a dynamic framework nowadays, leading to welfare-loss caused by congestion problems, arguing in favor of the possibility of differentiation of data packets according to their quality sensitivity [1].

EU urged that all traffic be treated equally but allow the network providers to make reasonable traffic management in consideration of justified technical requirements, so as to preserve the security and integrity of the network or to minimize temporary or exceptional network congestion. According to FCC rules, the no-blocking rule, the no-throttling rule, and the no-unreasonable interference/disadvantage standard will be subject to reasonable network management for both fixed and mobile providers of broadband Internet access service. Figure 8.3 shows that EU comparison with the United States rules. We can see that regulators paid attention to the innovation when making their

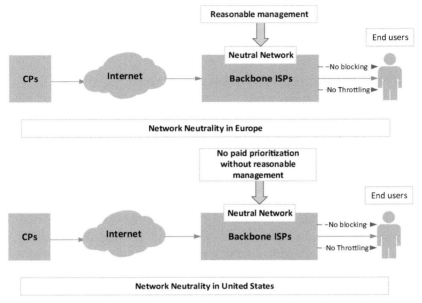

Figure 8.3   The comparison of NN between the US and the EU [8].

own NN policies. They leave room for the innovative services together with strict constraints.

8.3 CONASENSE [3]

In the past 10 years, wireless and Internet technology and system have created an explosive growth in personal and group applications which offered a wide range of data services. The limited data transmission capacity, as the bottleneck in the beginning, was broken. We can get much higher data rates over the same wireless channel and thus for support of many more demanding services. At the same time, we observe a rapidly increasing demand and innovative application areas for services related to positioning, tracking and navigation of some users/platforms. Similarly, an unprecedented development in sensing technology, sensors and sensor networks is being observed. A variety of sensors types are now available on the market in many domains. The integrated provision of these services will obviously raise people's living efficiency and Quality of Life (QoL). But traditional approaches may not be optimal, because of the allocation of different frequency bands, waveforms and hence different receiver platforms for these services.

Most promising CONASENSE services should be available in 5–20 years. The new CONASENSE services should reflect the trend toward an information society in which applications and services become equally important, bearing in mind that computing and communications should be integrated so as to save energy, software defined radio combined with cognitive radio concepts become increasingly important for new developments.

To achieve this goal, it is critical to identify the requirements for energy, terminal/platform and receiver/system design concerning diverse application areas including e-health, security/emergency services, traffic management and control, environmental monitoring and protection, and smart power grid. Minimization of energy consumption and energy harvesting deserve special attention in the novel CONASENSE architecture design mainly because of requirements for mobility, high data rate communications and signal processing and green communications. The novel CONASENSE architecture design will address problems and drawbacks of the existing infrastructures/architectures and be sufficiently flexible for future/potential developments. Consequently, the design of the CONASENSE architecture will be carried out so as not only to integrate existing and novel communications, navigation and sensing services but also to provide smooth transition between existing and new systems in hardware and software. NN impact on CONASENSE service innovation will be discussed in the next section.

8.4 Network Neutrality Impact on CONASENSE Service Innovation

From technology aspects, no discrimination requirement of Network Neutrality may have some negative impact on the new service innovation. The network nowadays is not neutral (Differentiated Network), it can supply quality of service (QoS) to different applications according to their characters and requirements. But according to the new NN rules, QoS measures will be taken as discrimination, and be banned in the pure neutral network, as shown in Figure 8.4. It is obvious to lower the network efficiency and will cause congestion easily.

In the short term run, NN will certainly make Internet content companies have more innovation space, encourage more innovative applications and increase the efficiency of the society. But in the long run, NN will inevitably weaken the enthusiasm of the investment on the network construction, lower

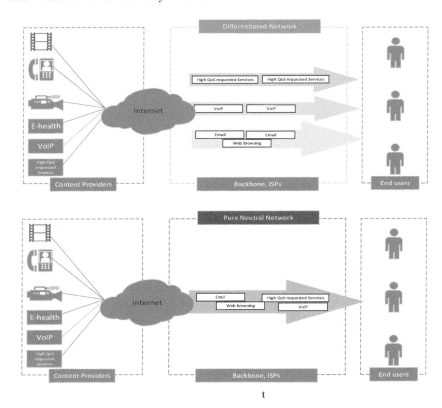

t

Figure 8.4   The difference between Differentiated Network and pure Neutral Network.

the network quality level gradually. This in turn will influence Internet companies in the end. The reasons are as follows:

(1) From the technical aspect, NN will impact the quality of the Internet service because NN limits the ability to guarantee the network QoS. As of now, there are mainly two types of model of QoS [12].

- Integrated Services (IntServ) uses RSVP (Resource Reservation Protocol) for signaling to invoke a pre-reservation network resource and traffic handling. IntServ can provide end to end guarantee for services and applications. But because Intserv is expensive and time-consuming, it has not been widely used in the Internet.
- Differentiated service (Diffserv) is a mechanism that identifies and classifies traffic in order to determine the appropriate traffic handling

mechanism. It can integrate the same type of services and manage them together. It is now used widely.

To exemplify the models of QoS, we can imagine an accident site, there are several wounded persons: some of them got severely injured and need to go to hospital immediately, while others are not so urgent. When the ambulances come, the doctors do not do any diagnosis and ask the severely wounded to get on the ambulance which makes some non-significantly wounded cannot get on the ambulance in time. This case is Best-Effort Model (no QoS). The IntServ model is that the wounded whoever is serious or not need to reserve the ambulance, if the severely wounded didn't make a reservation, he will not be sent to hospital. In the Diffserv model when the ambulances come, the doctors will distinguish the level of wounded and ask serious wounded persons with similar situation to get on and be treated. So we can find Diffserv is an effective method for application and service transmission on Internet. But with NN, differentiated service is banned in public network.

(2) Different Internet services and applications have different requests. QoS uses four parameters to judge the services request including bandwidth, delay, delay variation and packet loss rate [13]. Table 8.1 presents the definition and the impact of the four parameters of QoS.

Table 8.2 shows that some services are sensitive to delay but not to packet loss, and some services are sensitive to packet loss but not to delay. It is necessary for a network to treat packets belonging to different applications differently in terms of their different requests. For instance, a network should give low-delay service to VoIP packets, but best-efforts service to e-mail packets [14]. Different services also need different QoS provisioning.

Table 8.1 Definition and impact of the four QoS parameters [13]

Parameter	Definition	Impact on QoS
Bandwidth	The maximum data that can be transmitted per second in the network.	It is to measure the transmission capacity of the network
Delay	The transmission time a service packet takes from one nodes to the other one	If the delay is too long, it will lower the QoS
Delay variation	It means the variation of the packets delays in the same flows	It is a key factor impacting QoS
Packet loss rate	It means the rate of the data packets that lost in transmission	Low packet loss rate will not impact the QoS

Table 8.2   QoS parameters for some applications [14]

Type	Bandwidth	Delay	Delay Variation	Packet Loss Rate
Email	low	Not sensitive	Not sensitive	Not sensitive
WhatsAPP	Medium	Medium sensitive	Medium sensitive	Not sensitive
Video application	High	Sensitive	Sensitive	Sensitive
E-commerce	Medium	Sensitive	Sensitive	Sensitive
IoT, Industry 4.0	High	Sensitive	Sensitive	Sensitive
CONASENSE services	Large	Sensitive	Sensitive	May not very sensitive

8.4.1 From Social and Economic Aspect

NN may reduce the incentive of investment on the network construction. As we know, the funds usually flow to the market with big profits. If there is not enough profit from telecom network, investment on the network will certainly be reduced. Internet companies may also lose their interest to improve the efficiency of transmission, because they do not need to pay for the overuse of bandwidth. Insufficient investment on network with the overuse of bandwidth will certainly result in network congestion and inefficiency.

8.4.2 The Influence of NN Abolition

The abolition of NN in US will have an impact on the telecom and Internet industries respectively. Firstly, the abolition benefits the telecom industry. The telecom operators (broadband service providers) will see significant positive changes. The enthusiasm for broadband infrastructure investment is expected to be significantly improved, and the "digital divide" in remote areas is expected to be resolved again. Specific to Comcast, Verizon and AT&T and other traditional telecom giants' rights and interests, on one hand, the ISPs open up a separate "fast lane" in the network, according to the quality of transmission services for the Internet companies, charge varying levels of transmission costs. On the other hand, the priority delivery of its own content business, such as Comcast can give priority to the delivery of its own media group -NBC's business content. For specific content, specific software and services, telecom operators have the right to limit.

Followed by the Internet industry will be polarized. After the abolition of the neutral network policy, the U.S. Internet industry expressed its protest to the FCC in its view that abolishing network neutrality is equivalent to giving telecom operators the power to manipulate Internet traffic and discriminating against Internet businesses, which will eventually aggravate the polarization of the Internet industry. First of all, for Internet giants such as Google Facebook and Microsoft, they will be forced to pay the telecom companies what they call "fast track" fees and buy high-quality services. However, due to the huge financial resources of the Internet giant, it is expected that there will not be too many negative effects. Second, for the small Internet businesses, the network transmission quality of startups may decline due to the lack of protection of network-neutral policies, transmission costs may increase, and the long run will affect network innovation and business growth. Precisely because of this, the US neutral network supporters that: "FCC's abolition of network neutrality policy, not to destroy Google, but the next Google."

The NN legislation debate in the United States has lasted for more than 20 years. The difficulties and hiccups among them can be imagined. The adjustment of this network neutral policy is, in essence, the rebalancing of the future development space between the telecommunications and internet industries by the U.S. telecommunications regulatory agencies. During the Obama administration, telecom companies invested a great deal in advancing the network and Internet companies enjoyed too much dividends. After taking office, Trump passed a neutral policy adjustment and handed over a part of bonuses to telecommunications companies. The FCC's stance reversal is compensatory from a deeper point of view.

In the United States, network neutrality will still be the focus of controversy. Comcast, Verizon and Google, YouTube fiercely competing with each other, but on the other hand, the two sides can take a cooperative approach, the Internet companies own innovation and development is the need for telecom operators to support In the future, the cooperation between the two industries is still greater than the competition. In the long run, the surge in broadband network traffic, the consumption of operators such as video and other services, the competition between basic operators and Internet companies, the upgrading of networks and the sustainable development of the market also exist. In addition, taking into account the further integration of telecommunications and the Internet, although there will be new opportunities for development in this process, conflicts are unavoidable in the process of concrete integration.

8.4.3 Impact on Industry

For telecommunications companies, it is undoubtedly a great benefit. As the network neutrality policy is abolished, there is no limit to prohibiting payment priorities. Broadband service providers will be able to offer differentiated treatment and charge Internet companies more money by opening up the network fast lane. From the perspective of the development of the telecommunications industry, the abolition of the network neutrality policy will help to improve the enthusiasm of telecommunications enterprise infrastructure construction and further improve service quality. In addition, some network neutral supporters have pointed out that without network neutrality restrictions, broadband service providers will prioritize their content and lower the priority of competitor content.

For Internet giants and small and medium-sized Internet companies, they will face different fates. For the Internet giant, on the one hand, the abolition of network neutrality regulations will enable telecommunications companies to access or limit the speed of users by charging more fees; on the other hand, these large content providers have funds and a large number of subscribers. Negotiating chips is conducive to winning the so-called fast track of the network and gaining competitive advantage. Therefore, for the Internet giants such as Google and Facebook, the negative impact is actually limited. But for small and medium-sized Internet companies, especially emerging Internet companies, the impact is huge. When network neutrality policies are abolished, telecommunications companies will probably require them to pay more for faster transmission speeds and quality of service, which will greatly increase business costs. Companies that oppose the abolition of NN further point out that once payment priorities are allowed, they may stifle emerging Internet companies because they do not have enough funds to achieve faster content delivery.

In addition, the abolition of the network neutrality policy will have a certain impact on consumers. On one hand, the payment-first approach will drive investment in broadband infrastructure, resulting in a much larger overall network transmission speed and better online experience. On the other hand, if consumers want to use a more fluent network, they may have to pay extra fees and increase user costs.

8.4.4 Impact on CONASENSE Services

The large amount of CONASENSE services, to some extent, can be taken as environment-sensitive services. And for people on air, on vessel, or on car, on road, their environmental situation including location, speed, temperature, health condition etc. are constantly changing. All the relevant information should be received correctly and timely for a CONASENSE services will probably use them to make a correction and in-time decision, to help improve people's Quality of Life (QoL). NN rules put relatively strict criteria of specialized services or no-BIAS services, sometimes on a case by case base. That may slow down the CONASENSE services experimenting, developing, commercializing process.

High data transmission capacity is the base for the CONASENSE services. Most of the CONASENSE services focus on the future, on the assumption of WISDOM or 6G network deployment and extensive use of sensors. The value of CONASENSE services will not be realized on a congestive and no guarantee networks.

8.5 Conclusions

NN rules, its relevant debates, and its impact on ISPs and ICPs are discussed in Section 8.2 as a background study. In this discussion, NN rules such as NB, NT, and NPP were covered including their exemplified definitions and impacts on present communication paradigm. It is discussed that these rules are user centric and may hamper the financial benefit of ISPs. The absence of any control in the flow of information through a communication network may refrain ISPs from earning quality based revenues. In such a case, ISPs may not be willing to enhance or upgrade the network to improve the QoS of the network. Therefore, although NN rules are beneficial for innovations, as campaigned by NN proponents, provided by NN rules, the end user may not "really" enjoy the QoS.

With the convergence of Internet and telecommunication network, basic telecommunication services have been moved to the public Internet network. Many of them has quite strict QoS requirement so as to guarantee the relevant service responsibilities, such as emergency call, the basic service quality level. In this process, the policy maker and regulator should be careful to handle between service innovation, lawful customer right and social responsibility. QoS models have enough reasons to remain as legal network functions. And currently, US has realized the problem, and decide to abolish NN.

In Section 8.4, it is discussed how NN rules, in their present form, will impact CONASENSE service in the future. Through some assumed futuristic scenarios it is discussed that NN rules may not be favorable for innovation service. Banning Diffserv may put innovation service in the category of a usual communication system and ignore the very high data demand of the CONASENSE service. As customers may not welcome the innovative but poorly served new technology, the aspirant companies will struggle capturing market for this innovative technology. Further, NPP may not allow users to choose better services among the choices offered by ISPs.

NN policy may stimulate the development of service innovation in short term but may be not good for the network base in long term, which may in turn hinder the service innovation, such as CONASENSE services. The relatively strict criteria for specialized services or no-BIAS services may slow down the CONASENSE services experimenting, developing, commercializing process.

Regulators are suggested to make careful decisions on NN policy to guarantee the long term benefit according to their own situation. All the stakeholders should be encouraged to find a way to ensure the prosperous development of the industry in the market on their own, especially in service innovative era [15].

References

[1] FCC, "The Open Internet Rules and Order" FCC 15-24, March, 2015.

[2] FCC, Category "For Consumers" "Open Internet", 2015.

[3] Cory Janssen, "Over the Top Application (OTT)," Techopedia http://www.techopedia.com/definition/29145/over-the-top-application-ott, 2015.

[4] Leo Ligthart, Ramjee Prasad, "Communication, Navigation, Sensing, and Services (CONASENSE)" River Publisher, 2014.

[5] Winston Maxwell, Mark Parsons, Michele Farquhar, Net Neutrality – A Global Debate, Hogan Lovells Global Media and Communications Quarterly 2015, P15–17.

[6] European Paralimentary "Our commitment to Net Neutrality", EU Actions, October, 2015.

[7] Tim Wu, Network Neutrality, Broadband Discrimination, Journal of Telecommunications and High Technology Law, Vol. 2, p. 141, 2003.

[8] FCC, "The Open Internet Order" FCC 10–201, December 21, 2010.

[9] European Commission, "Net Neutrality challenges", October, 27, 2015.

[10] Dong-Hee Shin & Tae-Yang Kim, "A Web of Stakeholders and Debates in the NetworkNeutrality Policy: A Case Study of Network Neutrality in Korea".

[11] European Commission, "Roaming charges and open Internet: questions and answers" 27 October 2015.

[12] El-Bahlul Fgee, Jason D. Kenney, William J. Phillips, William Robertson1 and S. Sivakumar, "Comparison of QoS performance between IPv6

QoS management model and IntServ and DiffServ QoS models" the 3rd Annual Communication Networks and Services Research Conference, 0-7695-2333-1/05, 2005, IEEE.

[13] Fabricio Carvalho de Gouveia and Thomas Magedanz "Quality of Service in Telecommunication networks", Telecommunication systems and technologies, Vol, II.

[14] Hua wei technologies co (2013), ltd, "QoS Technology White Paper", http://e.huawei.com/us/marketing-material/onLineView?MaterialID=%7B3623FE01-3572-4413-A71B-EBEBE9F2E141%7D.

[15] Ramjee Prasad, 5G Revolution Through WISDOM, Springer Science+Business Media New York 2015, Wireless Pers Commun (2015) 81: 1351–1357 2015 March.

9

Enabling Security through Digitalization and Business Model Innovation

Albena Mihovska[1], Peter Lindgren[2] and Sofoklis A. Kyriazakos[1]

[1]Department of Business Development and Technology, Aarhus University, Denmark
[2]Aarhus University, Denmark
E-mail: amihovska@btech.au.dk; peterli@btech.au.dk; sofoklis@btech.au.dk

The developments and innovations in the area of mobile wireless communication systems that we have witnessed over the last couple of decades have been motivated by the objective to deliver seamless user connectivity anytime, anywhere and anyhow. Digitalization is a natural product of achieving ubiquitous connectivity enabling seamless digital experience through all-around availability of various data, which in turn is a driver for business model innovation (BMI). This Chapter addresses two research questions. Firstly, it explores digitalization as business value co-creation catalyst. To this end, three case studies are presented. Secondly, it explores the security challenges and opportunities arising from a rapid digitalization. Finally, it presents an approach to enabling security in the digital ecosystem through BMI.

9.1 Introduction

Digitalization is a process that has the power to restructure many areas of the social and business life, focusing the activities around the communication and utilization of available digital information from a variety of sources. The progress in wireless sensor technologies, machine-to-machine communications, big data analytics and artificial intelligence, on the other hand, have created an opportunity for businesses to offer new services to their customers as well as to adopt more cost-effective solutions for their operations. The ability to select the right tools for the processing and analysis of the data obtained from the sensing environment and to make these data accessible to the end user in the form of user-friendly, however, mostly time-sensitive applications, is one technical aspect of enabling value generation from these new technologies. Innovation of the existing business models then is a way to capitalize on this value.

9.1.1 Enabling Technologies for Digitalization

Current rapid developments in wireless connectivity, such as ultra-reliable low latency communications, haptic user feedback, high throughput capacity, high user-experienced data rate, energy-efficiency, minimal connection downtime and minimal packet loss, have brought forward efficiency and flexibility capabilities of the wireless networks, which are essential for time-critical performance over a wireless connection. In addition, the adoption of software-defined networking (SDN), network functions virtualization (NFV) technology and information-centric networking, have allowed for the protection of collected, exchanged and transformed data by means of on-the-fly deployments/positioning of security/privacy functions.

The cloud has changed the way businesses can connect and communicate. Cloud computing has increasingly become an integral part of business enterprises, allowing not only for the reduction of the total cost of ownership but also for generating revenue streams from the data continuously collected from the sensing environment and enabling their storage and processing to deliver applications upon personalized needs. Cloud-based solutions allow companies to add tools to almost any existing or future infrastructure, and are also a first step in enabling to share securely essential data with stakeholders, customers, partners, or vendors. The cloud allows for data to be stored in one place, which increases data integrity because users are able to always access the latest and correct version of data. Applications are in the cloud and available and instantly accessible from any device, which allows for saving time and improving productivity.

With progress in the area of Internet of Things (IoT), also pushed forward by the advanced performance requirements of 5G technologies for faster and reliable communication between sensor nodes, the number of connected applications and devices continues to grow exponentially, generating more and more data of business value. Jointly, all the above technologies allow for the reliable continuous gathering, exchange and processing of data in support of real-time decision making in a given scenario. This is where most companies put their digital focus around because the use of new technologies has the potential to ensure availability of contextually rich data, translation of these data, and a better management of the company's assets [1].

In fact, the development and use of advanced digitalization capabilities combined with smart connectivity technologies, has allowed for the spur of numerous cross-sector applications with a potential to generate new business value proposition. Digital innovations in response to fundamental socio-economic challenges, have transformed traditional economy sectors into smart grid, e-health, intelligent transportation systems (ITS), mobile money, smart metering technologies for water, gas and electricity management, and smart factories [2].

The increasing generation, processing, storage, and use of large amounts of data offer a new business resource. The explosion of data generation, the access to open-source and online data, and the possibility to extract relevant context, has become possible due to advances in artificial intelligence (AI), that provide the required computational capabilities for performing complex calculations to be applied to predictive and real-time decision-making. AI techniques, such as machine learning, deep learning and natural language processing, have been applied across various industries and open the possibility to redefine products as an opportunity for servitization of customers [3]. More and more businesses have come to realize that AI has the potential to generate value from the data collected one place and their interpretation on various operational levels within the company.

A strong example of the above is the healthcare sector. AI may be applied to numerous areas within the healthcare sector, such as remote monitoring and treatments, the prevention and detection of diseases, clinical trial research, remote surgery, and personalized treatment, including general improvement of the quality of life [4–6]. Data drives all healthcare activities and many such activities are information management tasks. However, there are still many limitations to both healthcare data gathering and data mining that can result in the generation of unreliable data sets and thus, impact the safety of the user, negatively. Healthcare data can be collected from the immediate user sensing environment, which will be based on measurements and sensing from consumer or medical devices (e.g., blood pressure, heart rate, sports or motion activity, etc) or environmental sensors and location devices. These data can be seen as micro data, or data from uncertified data sources because they will originate from non-medical related organizations. These sensing data would carry vital correlations to the data collected from certified sources, such as electronic health records, patient-reported outcomes, data derived from Internet use, genomic information, medical records, clinical laboratory equipment and will be used to trigger alarms when deviations from a norm are detected. However, usually digital prescriptions and medical reports as entered by the physicians are narrative and unstructured and would require the use of text mining tools, which, however, are sources of errors. Typical health records contain few data points (for a single event and a single individual) and many features. This means that the statistical modelling through data mining methods becomes very challenging and the validity of the results is limited. One way to deal with this is through real-time analysis of large diverse and unstructured data sets, which are characteristic for healthcare data.

When we deal with high dimensional data (i.e., macro data), i.e. a large number of records and/or features, the complexity of the data mining algorithms has an impact on the duration of the data analysis. Further, in large datasets, it is not an easy task to find outliers (i.e., a record that is unlike most others in the data set), understand if they come from errors in the data collection process and eventually decide to discard them. Finally, the display representation of the analytics results in a meaningful format is challenging when a large number of features is considered, as it is for typical digital health records.

Recently, AI has increasingly been applied by manufacturing and utility companies for smarter preventive and predictive maintenance. In general, maintenance becomes increasingly important, the larger the amount of psychical assets a company owns.

Another area that has made good use of the ML technology is as prevention of fraud during online services and transaction. ML has recently been considered for converged use with blockchain for an improved security and interoperability due to the data analysis capabilities it brings into the process. The latter can allow for building better models that integrate the distributed ledgers to increase data sharing capabilities. Distributed ledger technology is a new technology and the full extent of its future uses is not openly visible. The ledger is a data structure consisting of an ordered list of transactions. Examples range from monetary transactions to goods exchanged between known parties, and, more general, to any exchange of data. The distributed ledger is, basically, a replicated data-structure, which can only be appended to. A system that supports distributed ledgers is characterized by its target applications (most popular being crypto-currencies, digital assets in general and general user-defined computations or smart contracts), by the number of ledgers and by the ledger ownership.

Distributed ledgers offer a range of benefits to government and other public/private organisations, as they can be distributed in a precisely controlled fashion. Changes by any participant with the necessary permission to modify the ledger are immediately reflected in all copies of the ledger, and, on the opposite side, they can be robust in rejecting changes by unauthorized parties, making the corruption of the ledger extremely difficult. This makes them suitable for avoiding the "single point of trust" problem.

9.1.2 Digitalization and Impact on Business Models

Any business model (BM) may offer a value proposition offered as tangible and/or intangible value [7–9]. Value proposition can be measured as products, services and/or processes of product and services. Value propositions may be offered as physical, digital or virtual values – or combined. A BM carries a customer and user dimension, where users as can be defined as not paying for the value proposition [10, 11], while customers pay for the value proposition [10]. Users can, however, "pay" with other value, other value transfers and, thereby, contribute to the development of very important values for other business models.

Digitalization brings in capabilities that have the potential to change current types of interaction and enable value co-creation with customers. Figure 9.1 shows how the digitization of product data enabled by IoT, advanced robotics brings about availability of data that can be used for more efficient production process, predictive maintenance, and involvement of the customers into the production process. These data, helped by cloud-based collaboration, advanced analytics techniques and integrated platforms allow for faster and flexible decision-making and business processes. Technologies such as virtualization, augmented and sensory reality helped by the collected data and their sophisticated contextual interpretation, allow for more immersive type of applications that lead to lower investment risks, enhanced with innovative services business portfolio and tailor made customer management.

It can be concluded from Figure 9.1 that the potential of digitalization of assets and goods to impact the AS IS BM, is determined by their 'digital-as-a-service' value. Value co-creation would be realized by the type of service the digital components would be able to provide, as well, as how many end users that service would be able to acquire.

The creation, capturing, delivering, receiving and consumption of value is enabled through relations; [8, 9, 12]. Relations connect the components of the different BM dimensions and enable the creation, capturing, delivering, receiving and consumption process of value. The challenge to future BMs and their innovation is the required speed of forming multitudes of relations necessary to run a future BM. Digitalization enables new types of business

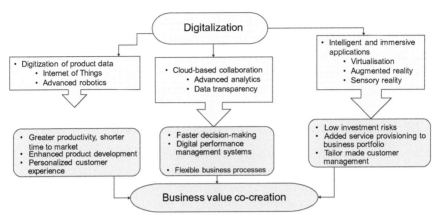

Figure 9.1 Digitalization as BM innovator.

relations but poses also a number of questions: What do we want to digitalize? Why do we want to digitalize it? How do we plan to digitalize it? What are the limitations, both, technical and operational to digitalization in terms of BM innovation?

Future BMs and BMI will be challenged on the collaboration forms, agility, flexibility, and virtual capabilities. Digitalization, is a key enabler of such attributes and a prerequisite for the multi-business model ecosystem (MBES): many new BMES will evolve and the industry, market and segment approach is expected to be disrupted with the business model ecosystem approach, which will enable much faster, agile and flexible BM and BMI and perform the BMI on all levels.

9.1.3 Security Challenges in the Digital Economy Era

Digitalization, and more specifically, the evolution it brings towards a connected, smart, and highly efficient business ecosystem, will change larger and smaller industrial players into platforms of business innovation. However, with a fast growing digital economy, trust and security have become a critical element [2].

With an ever-growing number of connected applications and devices, the guarantee of integrity and protection of the data these generate, will be key to the realization of economic value. Data integrity is directly related to the quality of the interaction, and, therefore, would impact the value co-creation process [3].

The ability to secure multi-business model innovation (MBMI) and to surround what is really taking place in an MBMI process or an operating BM – "AS IS BM" – has become a very hot topic in many businesses and societies these days. There are a variety of reasons to this security issue and many technological ways to provide security related to this matter. In this chapter our focus is mainly on four major issues.

Firstly, as business these day change from business mainly carried out between humans to business carried out between humans and technology and technologies to technologies – "machines to machines" – the security issues move from human-based security to technology-based security and, therefore, brings often new security challenges. With digitalization, humans would lose control and management of security to a large extent.

Secondly, as business and BM's becomes more and more network-based and with network partners that have to exchange business models between each other within seconds, security becomes an important issue related to how and when to be sure about business collaboration and exchange.

Thirdly, when network business models increase exponentially and no previous knowledge of each other network partners exist, the risk of doing business increase and trustfully security systems need to be developed.

Fourthly, the overall vision and purpose of technology and business modelling is to develop these as user friendly, seamless and with best possible interconnectivity – anytime, anywhere and anyhow. The challenge is how to do this so that security does not course friction in the BM process and does not interfere and have a negative effect on the users, customers, network partners and employees that the BM experience.

9.2 Case Studies for Value Co-Creation Through Digitalization

9.2.1 Healthcare

The digitalization of the healthcare system isunderpinned by the implementation and use of information and communication technology (ICT). The use of ICT in the healthcare system is defined as e-health and has potential benefits for optimizing the processes and for improving the health data collection by, for example, tracking persons and objects in the hospital or smart monitoring of patients by wearable technology [13].

An improved systematic use of health data with a focus on the management processes and the development of competencies is critical for an improved patient involvement [14]. Digitalization of data, procedures and workflows is part of the development and improvement of the healthcare system. The monitoring is a continuous observation of a patient's vital signs, and it is necessary for a correct diagnosis. The traditional way of monitoring is manually using technology in the form of devices that can monitor a specific vital sign of a patient. The device may be, for example, an electrocardiogram (ECG) for monitoring heart rhythms, a thermometer, for the measurement of a patient's temperature or inflatable pressure cuffs with a stethoscope to monitor the blood pressure. The monitored health data are assessed and analyzed by clinical personnel such as medical doctors and nurses.

9.2.1.1 Monitoring of patients

The technology of patient monitoring has advanced in the recent years to include and obtain data from wearable sensors and devices, which can continuously and in real-time monitor and send the patient's data to the clinical personnel [4, 5, 15]. The remote monitoring has the advantage of providing vital health data related to a patient independently of a specific location [15]. Monitoring is a vital part of the treatment and diagnosis of patients and it is one of the areas that are in rapid development with new technologies emerging, such as the IoT, smart technology and big data. The Danish healthcare system is, in general, a front-runner in the use of digital health and the system is characterized by its large use of electronic communication and IT systems in hospitals. National strategies for improvement of healthcare focus increasingly on the high use of digital health data in a combination for both primary- and secondary-user purposes [16]. The primary-user purpose is using digital data for the direct care and treatment of patients and citizens, while the latter is using digital data for research, quality assurance and management in the healthcare sector.

Health data are systematically collected throughout the Danish healthcare system, including in the hospitals and by the general practitioners. The large amount of data is an important part of the monitoring and treatment of patients and the data are saved in electronic health records (EHRs), the national patient register and medication databases among others. The health data are transported across, both, the primary and the secondary sector, as well as across several departments and institutions such as hospitals, general practitioners, local authorities and home care services.

The Danish healthcare system has a large focus on the prevalence of IT standards for facilitating such data transportation through electronic communication [17]. The Danish healthcare system's high use of ICT, digital workflows and e-health, in general, secures a fully integrated use of digital health data. However, the Danish healthcare system seeks to improve the quality and efficiency of the offered healthcare services with a coherent collaboration across sectors and departments, while having a high focus on continuously improving the use of digital health [17].

A wearable monitoring technology, which can remotely monitor the vital signs of patients is an example of use of smart technology and consists of objects that can exchange, store and process data and information applying the concept of IoT [18]. The possibilities of IoT in the healthcare industry fits the emerging challenge of an increasing number of patients with chronic and long-term diseases due to the fact that it can create value by providing different monitoring functions and application possibilities [4, 5, 18]. This is also supported by a rapid evolvement of wearable technology, such as computerized watches, sensors integrated in textiles and smart glasses. The trend is towards technology with minimized sizes and a higher possibility of monitoring biomedical data, including the vital signs of patients [19]. The use of wearable devices for monitoring is potentially making the treatment process more patient-centric hence making the individual patient possible of managing their own health data monitoring and giving the possibility of continuous ambulatory monitoring of vital signs with the advantages of enabling mobility and minimizing interference with other activities [20]. Normally, the introduction of digital technology for one purpose can easily enable additional value co-creation for other processes. A study performed by [21] showed an additional value of the implementation of DASH7 technology for a radio-frequency identification (RFID) tracking system that places tags on beds, materials and equipment and makes it possible to locate either inventory or patients in a hospital environment. Aarhus University hospital (AUH) in Denmark has currently (at the time of writing of this Chapter) the largest such installation in the world with 1.000 hospital beds and 300.000 pieces of material as part of the RFID tracking system with the availability of locating these items wirelessly [21] explored the use of this technology to enhance it with wireless patient monitoring and showed that patients can benefit from increased mobility, which is of particular importance for in the case of hospitalization of children, while medical staff would have their working load decreased without jeopardizing the patient's safety. However, the main challenges to the full realization of such a scenario are open issues in terms of a somewhat complex setup, reliability related to the localization of patients and obtained data quality.

9.2.1.2 Digitalizing TOKS

Monitoring of vital bio-signals of patients can be enabled by several different monitoring technologies and devices, stand-alone or integrated into a platform [4]. The signals can be bioelectrical or biochemical signs and the signals would be monitored, analyzed and assessed for diagnosis and the treatment of patients [4]. Applying digital technologies for improving the monitoring of patients, has also a good potential for making the treatment more cost-effective.

An example of enabling value by use of digital technology is the improvement of a standard process implemented in the hospitals of Region Nordjylland, Denmark, known as TOKS (early discovery of a critical illness) [22]. TOKS has been implemented because of a lack in a systematic observation of very important values of the patients and a lack of taking action when these values are deviant. This could result in non-planned hospitalization of patients, heart failure or in the worst case, death.

The state of the patients would be determined based on the vital values of the patient and would be measured by the medical staff (e.g., nurses) at the hospital. Related to TOKS, vital values are physical values and include: frequency of breathing, oxygen saturation, blood pressure, pulse, consciousness, temperature [23]. The idea behind using TOKS as a process is to measure and observe patients and their state systematically at least once every day. The TOKS objectives would be to identify patients whose state gets worse and to create an official filing of the observation and measurement of the patients' vital values. Values would be entered in the electronic patient records (an approach adopted in the Danish healthcare system since the year 2000), where the system will then calculate a score according to a decision algorithm. This algorithm would determine how critical the illness of the patient is, and, furthermore, determine how often observations and measurements should be taken on the patient, as well as it would indicate any other actions that should be taken [23].

The experiment reported in [22] showed that TOKS approximately would take five minutes by each patient, including all setup and measurements – and documenting the values through a smartphone. Blood pressure through the arm cuff takes approximately 40 seconds and using the Pulse Oxymeter takes approximately 20–30 seconds. These as well as walking back and forth between the patient and the stand with medical devices are procedures that take up time. Some values would be written down on paper, before entering them into the system with the smartphone later, so it was estimated that approximately 30 seconds were used on this process. Finally, some time

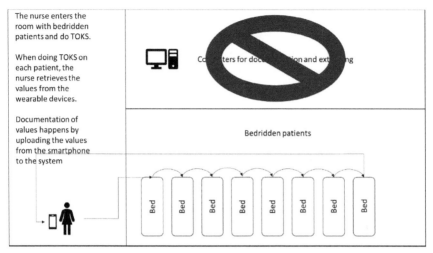

Figure 9.2 TOKS improvements for medical personnel by use of digital technology.

would be spent on greeting, small talking and getting information from the patients. This all together took one minute and 30 seconds. Greeting and small talking with the patients cannot be valued in time because many patients (elderly, in particular) would appreciate and need the face-to-face interaction in order to feel taken care of and safe. Overall, during the performed observations, TOKS would take approximately 55 minutes in a bedridden department with ten patients.

The time for extracting the data that the medical doctors would need to make decisions about the further treatment of the were not recorded. This was due to the independent situations of each patient. It is difficult to give a precise amount of time spent, but by implementing tablets by the beds, an assessment can be automatically visualized for each patient during the visit of the medical staff. Implementing the new TOKS would result in either one less employee hence fewer expenses to the hospital, or more time to employees, which is also a desired wish by the nurses in terms of stress and feeling pressure. A sketch of the improved TOKS is shown in Figure 9.2.

9.2.1.3 ICT-based rehabilitation

Rehabilitation is a common term for post treatment of patients that suffer from chronic disease. Providing a suitable rehabilitation is the main issue for chronic patients as it helps to live independently and enhance their quality of life. However, as the rehabilitation period usually lasts some months, the continuity of care often is interrupted in the transition from the hospital to the home. Virtual Coaches can help these patients to proceed with a personalized rehabilitation that complies with age-related conditions, as the key technology for empowering patients through the enhancement of the adherence to the care plan and the risk prevention.

Cardiac rehabilitation has been recommended as a part of the overall treatment for cardiac patients and is has been proven that it has positive effects on the morbidity and mortality in cardiac patients. It has been shown that telemedicine rehabilitation could be an effective and favorable expansion of the rehabilitation program [24]. A central e-Health platform that serves central infrastructure services and obtains the information delivered by sensors or gained by the direct interaction between the patient and a virtual coach was proposed by the funded under the European Framework Horizon 2020 program project vCare [25]. The high level architecture of the vCare platform is shown in Figure 9.3.

Devices can be added to this platform, such as camera, microphones and a Kinect sensor, which makes the platform able to track movements. The information from the devices would be conducted by a real-time processor. Besides the platform, the infrastructure delivers supporting services to improve the quality of life of patients. The service provides physical and cognitive exercises as well as educational material about nutrition and healthy life behavior. In vCare, this service has been extended with a care pathway

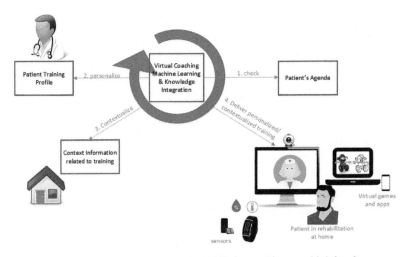

Figure 9.3   VCare proposed rehabilation architecture-high level.

and a knowledge layer that enables personalized exercises and material for the given patient. Based on the use of reasoning algorithms, the virtual coach is foreseen with flexibility regarding the patients' needs and, hereby, able to make specific rehabilitation programs. It is possible to implement the platform on different devices, e.g., tablets, smartphones, TV screens, and so forth [25].

This modularization opens the opportunity to enhance the functionality that vCare can provide for the rehabilitation process of patients. Services and applications provide the main functionality of the vCare platform. Every utilized service aims at addressing different problems and tasks in order to support the patients and medical experts in their needs and goals. The vCare services can be distinguished by two different types: back-end (business logic) services that provide basic and required functionality (e.g. monitoring, reasoning, activity recognition) and additional (front-end) services that depend on additional and individual requirements of the target users and support the user interaction with the virtual coach. On the one hand, the vCare service platform aims at integrating and re-using already existing successfully applied services in the healthcare and ambient-assisted living (AAL) domain, on the other hand, it has the objective to facilitate the integration of new developed vCare services. The integrated and applied services mostly depend on given medical cases and have to be defined and decided by medical experts, considering the patients' characteristics and their medical records.

[24] Investigated the cost-effectiveness of remote rehabilitation within the Danish healthcare system and showed that a combination of ICT-based and standard cardiac rehabilitation could improve quality of life, while offering significant cost-benefits to the healthcare providers.

9.2.2 Smart Metering Solutions

Energy industry, specifically the heating sector, has shown much potential regarding digitalization and implementation of new technologies [26]. Smart heat meters when installed in a heating distribution grid provide enough data to detect disturbances and issues in the network very fast and provide many opportunities for both heating utilities and metering solutions manufacturer. Digital platforms have been proposed, whose purpose is to continuously visualize data obtained from the smart heating meters of the heating utility's network and inform utilities of i.e. a pipe breakage or heat losses occurring in the grid [26]. The need for heat loss and leakage surveillance, need for continuous optimization of heat distribution grids and heat plants, process monitoring and similar, are further motivators for employing digital technologies in this sector.

The expansion of the business model around a digital service platform will require higher level of data and analytics capabilities. This causes that the line between customers and providers starts to fade, and the customers can give valuable knowledge to the providers and vice versa, which would drive the functionalities needed to provide intelligent digital services, which in turn is a co-creation of value [26]. This would have an impact on the revenue streams to change from traditional asset-selling to more innovative and experimental models.

9.2.3 Agriculture

Agricultural activities over the globe have been limited by natural resources, financial support, limitations of the existing information systems, and lack of trust in sustainable agricultural methods versus non-sustainable methods. The recent trend of preferring farm products produced by organic methods over products based on the excessive usage of chemical fertilizers and pesticides, have given impetus to sharing of the knowledge and resources of farming amongst well organized and/or well informed agriculturists.

Agriculture analytics that can reduce the producing cost, prodigal use of government investment, minimize the cost for storage and transport logistic, and enhance marketing at the right place, among others.

Knowledge creation and accessible information systems and tools to monitor, gather, transform and above all share vital information between key stakeholders can help the sector to become more sustainable. However, as well as the potential for new knowledge, a substantial part of the existing knowledge and its underpinning information flows, has yet to be exploited to its full potential.

The recent trend has been towards integrated digital platforms for agriculture that by automating data ingestion and analysis can provide for value creation. Integration enables and protects stakeholder access and information; automates the development and analysis of massive bodies of data; and develops, reveals, and manages the potential costs and revenues of these decisions. Automation is a key enabler here because it manages the analytic processes applied to every file and every data element of that file. It analyzes the data to support some anticipated event and its required decision. These data exist in very large files, and these files would be transmitted across available communication systems. These files must be in forms that are accessible and actionable, able to be stored and analyzed in a timely, as-needed, and cost-effective way. Stakeholders should have the ability to progress from data scheduling, presentation, and description to developing predictions, diagnostics, and prescriptions for action. An integrated digital platform applies the analytics and adds knowledge to the process; it must traverse the data from what happened, to why it happened, and how it can be mitigated.

Smart farming involves massive volumes of multi-source data that can be registered, interpreted and used in the decision-making process. There are numerous application scenarios where data mining and ML techniques can improve the farming process (e.g., predictive analysis in farming operations, real-time decision-making, collaborative systems in food supply chain networks) [27].

9.3 Digitialization, Cybersecurity and Business Model Innovation

Although companies are rapidly adopting digital technologies for improving their operation management and processes, there are many challenges to digitalization and, more importantly, to capitalizing on the opportunities it presents. The fact that digitalization is basically about the continuous collection of data puts the challenge of cybersecurity, and how to guarantee such, a major hurdle to overcome.

Fault tolerance and business continuity are very important for digital service platforms to ensure that service recovery (restarting the services) and rerouting of requests for services are automated. Security is a key requirement to ensure that personal data or data pertaining to third-party users are secure – data encryption, compulsory user-authentication, 2-step user authentication for access to critical data can be considered.

Compliance requirements also need to be considered for some services – though most of the compliance issues might get covered in data security, however, some compliance issues need to be covered if the services deal with financial systems.

Digitalization inevitably integrates and makes the various BMES interact, which is an opportunity for integrating security, privacy and trust as an embedded feature of the digital BMES. Based on already introduced advanced techniques, such as ML, automation, data processing algorithms, and so forth, simply for capitalizing on the context carried by collected and open-source data, allows for optimizing the digital algorithms with sophisticated cybersecurity-related features. This has the potential to evolve the businesses embracing digitalization in the scope of a current portfolio of products or services to a new role of being their own security providers, and possibly security providers for other players of the BMES.

Because in the digital BMES, the data are shared, establishing trust among all players will be a key factor for co-value generation and for sustainability of the business relations. In essence, the BM of the future will bridge various security technologies that would integrate together to self-design an appropriate security strategy, which will be characteristic for each BMES.

The BMES digital capability represents the ability to configure hardware components to sense and capture information with low human intervention. The components that enable the above digital capability bring into the BMES intelligent functionalities. This in turn releases new possibilities to inject intelligence into dealing with security and trust challenges.

9.4 Conclusions

With the recent advent of big data analytics and ML that enable optimized decision-making systems, the ICT industry offers a range of solutions addressing the need for digital transformation to all actors involved in global value networks of many industries and markets worldwide. It is important to account for the fact that with all its positive potential, digitalization brings about increased sources of digital waste and their proper management is directly related sustainable security.

When we digitalize processes and roles that make up the operations of a business, we digitally transform the business and its strategy. Predictive analytics, AI, blockchain, big data-related technologies are technologies that carry this potential. Blockchain is still in the initial state of research but its potential to unlock the full potential of digital capabilities in may application areas has been recognized by researchers, strategists and standardization.

Another key prerequisite for generating novel business applications is the ability to generate and use high quality data, which can be achieved by a capability to merge data from various data streams.

Distributed ledgers can enable novel rich business applications by guaranteeing interoperability between different applications. An architecture that is based on a distributed ledger does not require that nodes are the same, only that the nodes of the system can be identified securely (i.e. has a public/private key pair). Distributed ledgers are essential to providing security by enabling authentication, confidentiality, accountability and traceability of application related data exchange, products and applications.

The power is in finding efficient ways in combining to the best advantage of a business all the above technologies together and this can be realized in the most powerful way jointly with the process of business model innovation.

Acknowledgement

The case studies presented in this Chapter have been based on student project work carried out at the Department of Business Development and Technology (BTECH) at Aarhus University, Denmark under the Technology-based Business Development Master program and on research activities within the EU funded H2020 project VCare.

References

[1] A. R. Jensby, Machine Learning and Predictive Maintenance Model. Project Report: Technology Specialization 2, Aarhus University, Denmark, 2018.

[2] Florent Frederix, Towards a European ICT security certification framework. ETSI Security Week, 2017.

[3] S. Lenka, V. Parida and J. Wincent, Digitalization capabilities as enablers of value co-Creation in servitizing firms. *Psychology and Marketing*, 34(1): 92–100, January 2017.

[4] S. Kyriazakos, et al., eWALL: An Open-Source Cloud-Based eHealth Platform for Creating Home Caring Environments for Older Adults Living with Chronic Diseases or Frailty. *Springer Journal on Wireless Personal Communications*, 97: 1835–1845. 2017. https://doi.org/10.1007/s11277-017-4656-7.

[5] A. Mihovska, et al., Integration of sensing devices and the cloud for innovative e-Health applications. Chapter 11 in *Wearable Technologies and Wireless Body Sensor Networks for Healthcare*, F. J. Velez and F. Derogarian Miyandoab, Eds, IET Publications, 2019, April 2019.

[6] A. Pnevmatikakis, et al., Game and multisensory driven ecosystem for an active lifestyle. *Proceedings of 4th FABULOUS conference*, Springer Publishers, 2019.

[7] P. Lindgren and Y. Taran, A futuristic outlook on business models and business model innovation in a future green society. *Journal of Green Engineering*, 229–239. River Publishers, 2011.

[8] Lindgreen, A. and Wynstra, F. (2005). "Value in Business Markets: What do we Know? Where are we Going? *Industrial Management*; 34: 732–748.

[9] P. Lindgren and O. H. Rasmussen, The Business Model Cube. *Journal of Multi Business Model Innovation and Technology*, 135–182. ©2013 River Publishers.

[10] P. Kotler, G. Amstrong, J. Saunders, and V. Wong, Principles of Marketing. Prentice-Hall Canada, 1999.

[11] E. Von Hippel, Democratizing Innovation, MIT Press, ISBN: 9780262720472.

[12] R.J. Brodie, R.W. Brookes, and N.E. Coviello, Relationship marketing in customer markets. in *Blois, K. (Ed.), The Oxford Textbook on Marketing*, Oxford University Press, Oxford, pp. 517–33, 2000.

[13] The National eHealth Authority, Making e-Health work. Technical report, The Danish Government, Local Government Denmark and Danish Regions, 2013.

[14] Ministry of Health, Nationalt kvalitetsprogram for sundhedsområdet 2015–2018, Technical report, Ministry of Health, Denmark, 2015

[15] M. Omoogun, V. Ramsurrun, S. Guness, P. Seeam, X. Bellekens, and A. Seeam. Critical patient ehealth monitoring system using wearable sensors. in Proc. of 1st IEEE International Conference on Next Generation Computing Applications (NextComp), pp. 169–174, 2017.

[16] Healthcare Denmark [2018], 'Danish digital health strategy 2018–2022 now available in english'. [Online; 2018.12.29] URL: https://www.health caredenmark.dk/news/listnews/danish-digital-health-strategy-2018-2022-now-available-in-english/.

[17] Ministry of Health, Healthcare in Denmark – an Overview. Technical report, Ministry of Health, Denmark. 2017.

[18] A. Yearp, D. Newell, P. Davies, R. Wade, and R. Sahandi, Wireless remote patient monitoring system: Effects of interference. in *Proc. of 10th IEEE International Conference on Innovative Mobile and Internet Services in Ubiquitous Computing (IMIS)*. pp. 367–370, 2016.

[19] M. Haghi, K. Thurow, and R. Stoll, Wearable Devices in Medical Internet of Things: Scientific Research and Commercially Available Devices. *Healthcare informatics research* 23(1), 4–15. 2016.

[20] D. Dias, and J. Paulo Silva Cunha, Wearable health devices–vital sign moni-toring, systems and technologies. *Sensors*, 18(8), 2414, 2018.

[21] A. B. Andersen and A. Mihovska, Wireless Smart Monitoring of Patient Health Data in a Hospital Setup. In *Proc. of 4th FABULOUS conference*, Springer Publishers, 2019.

[22] J. O. Poulsen, Digitalizing TOKS in the Danish healthcare system. Project Report: Technology Specialization 1, Aarhus University, Denmark, 2019.

[23] D. Trabjerg, Tidlig opsporing af kritisk sygdom – TOKS. 2018. [Online] Available at: https://pri.rn.dk/Sider/10881.aspx.

[24] M. B. Andersen and L. S. Larsen, Rehabilitation Strategies for Patients with Cardiovascular Disease. Project Report: Technology Specialization 1, Aarhus University, Denmark, 2018.

[25] H2020 EU-funded project Virtual Coaching Activities for Rehabilitation in Elderly-VCare, [Online], https://vcare-project.eu/.

[26] E. Dzanic, Digital business model innovation – case of a metering solutions manufacturer. MSc in Engineering – Technology Based Business Development, Report, Aarhus University, Denmark, 2019.

[27] S. Wolfert, L. Ge, C. Verdouw, M.-J. Bogaardt, Big Data in Smart Farming – A review. *Agricultural Systems*, (153): 69–80, ISSN 0308-521X, https://doi.org/10.1016/j.agsy.2017.01.023. 2017.

10

Secure Positioning and Navigation with GNSS

Nicola Laurenti and Silvia Ceccato

Department of Information Engineering, University of Padova,
35131 Padova, Italy
E-mail: nil@dei.unipd.it; ceccatos@dei.unipd.it

As our societies have come to increasingly rely on global navigation satellite systems (GNSS) for many critical tasks and the control of crucial infrastructures, so have increased the incentives to attack them for financial, or political motives. Two kinds of threats are particularly relevant to the GNSS context: *jamming*, carried out at the physical layer and aiming at denying the service to receivers in a given area, and *spoofing*, that aims to induce the receiver into computing a false position, velocity or time, by forging or illegitimately modifying the GNSS signal either at the data/message level or at the code/signal level. In this chapter we review the several countermeasures that have been devised, especially against the spoofing threat, both at the data and signal level, and discuss their effectiveness in relationship to the context and the possible side information available to the attacker and receiver. Eventually, we propose novel models for the analysis and design of partial spreading code encryption and for the integration of side information coming from other sensors into the integrity verification for the computed GNSS trajectory.

10.1 Basics of GNSS Positioning and Navigation

GNSS systems provide accurate, continuous, worldwide, three-dimensional position, velocity and time information to users with the appropriate receiving equipment. While the GPS system has been part of many people's daily

Security within CONASENSE Paragon, 145–166.

experience worldwide for decades, there are currently several other GNSS constellations providing analogous services, such as Galileo, GLONASS, BeiDou (24 to 33 satellites per constellation) and QZSS (4 satellites for regional coverage). GNSS constellations differ for geometry and their satellites are located in Medium Earth Orbits (MEO), ranging from 19,000 to 35,000 km altitude.

The transmitted GNSS signals differ among the several systems, but they all include a ranging signal and a data dissemination channel, that incorporates information about the satellites position, time corrections and several other parameters that are useful for positioning. GNSS can serve an unlimited number of users, as the network works only in downlink and receivers operate passively, calculating their position through the concept of one-way time of arrival ranging. The central frequency for these signals is around 1.5 GHz, but usually each satellite can transmit more than one signal at different frequencies (e.g., GPS L1 at 1,575.42 MHz and L2 at 1,227.6 MHz). The separation between signals at the same frequencies from different satellites is achieved through Code Division Multiple Access (CDMA), in which each satellite transmits a different pseudorandom code (PRN). The PRN codes were selected in order to have low cross-correlation properties with respect to one another. GLONASS, however, exploits Frequency Division Multiple Access (FDMA) rather than CDMA.

Users determine their position by measuring the propagation time of the signals from at least four distinct satellites (3D position and time). Converting the propagation time to distances and knowing the position of each satellite it is possible to perform multilateration and obtain the receiver's position, as shown in Figure 10.1.

In order to compute the travel time from a satellite to the user, the receiver utilizes its own clock (synchronized with the GNSS time with some possibly time-varying bias) and the publicly available PRN sequence. While resembling a random sequence, the PRN is indeed deterministic and known to all users. The receiver detects the signal time of arrival by generating its own local replica of the PRN and evaluating the correlation between the incoming signal and the local replica. The highest peak in the correlation indicates the estimated time of arrival for the signal and allows to calculate what is called *pseudorange*: the distance between user and satellite plus a residual bias due to several error sources (time synchronization error, ionospheric and tropospheric effects, multipath, etc.).

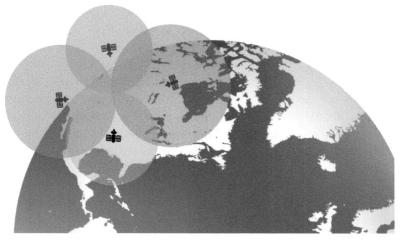

Figure 10.1 The principle of multilateration for the computation of the receiver position, with GNSS signals.

10.2 Threats and Attack Scenarios

10.2.1 Application Contexts for Secure GNSS Positioning

GNSS delivers global position, velocity and timing all over the world and is widely used in various commercial applications [1]. Transportation is one of the major areas in which GNSS is used in combination with other positioning systems to track the real time movement of vehicles. In railway applications GNSS is part of the monitoring system and is used to keep track of the position of locomotives, rail cars and maintenance vehicles; moreover it has a fundamental role in safety enhancement. Aviation exploits GNSS in collision avoidance mechanisms and relies on it for en route through precision approach phases of flight [2]. In maritime applications GNSS is necessary for the navigation in hazardous waters as well as for aiding maneuvers in congested ports, for mapping, underwater surveying, and more. Both the construction and the mining industry benefit from GNSS, which allows to obtain a higher level of automation and efficiency. In fact, GNSS is integrated into equipment such as graders, bulldozer, excavators, drilling and farm machinery for achieving higher precision and minimizing costs of operations while reducing the number of necessary on-site surveys. The use of unmanned vehicles, remotely guided by GNSS based applications, is no longer limited to the defense industry. The market of unmanned vehicles has grown and diversified and it has spread to other areas such as search and rescue, disaster management, surveying, infrastructure inspection and environment research.

Positioning is not the only service provided by GNSS. Several distributed systems rely on GNSS for precise timing information, as satellites are equipped with atomic clocks: telecommunication networks, financial transaction management systems and electric power grids are some examples of the many applications that achieve time synchronization through GNSS signals. As GNSS are now fully integrated into security critical commercial application, the potential impact of their being compromised would be rather severe. Sometimes systems are attacked with the sole aim of disrupting services, impacting the population or causing damage to institutions that rely on them. However some of the more recent GNSS applications may provide further incentive to malicious behavior, since a service disruption may directly lead to economic advantage. This is the case for road tolling or parking management applications, for instance, where GNSS receivers are used to monitor the vehicles' movement and apply service charges. Direct benefit may also come from hacking geo-fencing mechanisms, where vehicles are monitored through their own GNSS receiver to verify that they do not traverse access-restricted areas (e.g., fishing vessels, no-fly zones, etc.). In these kind of scenarios the attacker becomes the user itself, while the target is the central monitoring system.

10.2.2 Location Based Services

The integration of GNSS with our everyday life has been further boosted by mobile phones and the dramatic increase of new smartphone applications relying on *Location Based Services* access policies. The widespread adoption of Location Based Services (LBS) started in the early 2000s after the discontinuation of GPS selective availability, that used to limit the GPS localization accuracy for civil applications [3]. Since then, the demand for robust, available and precise localization techniques has increased for both indoor and outdoor applications. As smartphones have become ubiquitous in our everyday life, more and more mobile applications have exploited the capability of interacting with the environment through Location Based Services.

Location based social networks (e.g., Foursquare, Instagram, Twitter, Facebook, etc.) enable users to upload geotagged content, share their location and be notified when friends are nearby, rate services and stores in their proximity and receive suggestions and advertisements tailored to their location. User uploaded content can in turn be exploited for research and management applications (e.g., crowd mobility modeling and real time event and traffic management). Location based gaming interacts with users' position and motivates them to traverse specific locations to collect objects and gain benefits in a virtual world (e.g., Pokemon Go) or even in the real one (e.g., geocaching).

Social networking and gamification are exploited by LBS for promoting healthy lifestyle, as in fitness tracking applications. Complemented with other sensors such as Inertial Measurement Units, location awareness allows to provide a remote health monitoring service to those in need (dementia suffering patients, fall detector for the elderly, etc.). Assistive applications have been developed to support visually impaired people in their everyday tasks (e.g., obstacle detection, personalized way-finding) and improve their quality of life.

Navigation is by far the most popular LBS application, offering guidance in unfamiliar environments and improving the overall traveling experience thanks to real time traffic monitoring, vehicle management, wayfinding assistance and road information (Google Maps, Wayz, etc.). It has also enabled the successful spreading of car and taxi sharing platforms (e.g., Uber, Lyft, BlaBlaCar, etc.) offering a real time monitoring service for both providers and clients and providing flexible and competitive services [4]. LBS have recently extended to several other fields, such as emergency and disaster management, insurance and financial applications, production process support, etc.

As these services permeate people's everyday life and become involved with critical aspects of our society, research has started questioning the implications of this phenomenon. In [5] the social impact of LBS on society is categorized into four main themes, namely control, trust, privacy and security. Since privacy and security tend to be in contrast, the authors of [5] underline the importance of an adequate regulatory framework. The security aspect of LBS regards ensuring the availability, authenticity and integrity of the exploited positioning information against environment related malfunctioning or malicious behavior by third parties or even by the user itself.

The security concern is often put aside when dealing with LBS on mobile phones, as the presence of an attacker capable of tampering with the position of other users seems far-fetched and unlikely to have a critical impact. For instance, it is generally believed impractical to spoof a vehicle's positioning in urban navigation scenarios on the grounds that the driver may easily detect the inconsistency between the spoofed position and the outside environment. As a matter of fact, the feasibility of such an attack was investigated in [4], where the adversary targets a user that is following Google Maps navigation, and aims at luring the car to a selected location, without being detected by the user.

The attacker's hardware for producing the GNSS spoofing signal is assumed to be located either in the same car or in a close one, and it can be remotely controlled and operated. In order to maintain consistency with the outside environment, the attack strategy searches the map for areas that mimic the road shape in the user's location, producing a spoofing signal that will suddenly move the user's position. The navigation app will then adjust the suggested path, as predicted by the attacker and guide the user to a different destination. As the road geometry is consistent to that visualized on the application, it is likely that the user will not notice the deception right away, as demonstrated through an experiment campaign in [4].

10.2.3 Jamming

Jamming is a denial of service threat that operates at the physical layer by superimposing a powerful enough adversarial signal to the legitimate one so as to render it impossible to track/acquire/decode by the receiver. As such it is often a preliminary step to more sophisticated spoofing attacks, since a receiver that is not tracking the legitimate signal is more vulnerable and prone to track a spoofed signal

Traditional GNSS jamming relies on transmitting high power signals with a wide distribution both in time and frequency. Several waveforms have been proposed to increase its power effciency, such as short time pulses, chirp, random frequency hopping, and so on. However, as long as they are concentrated in some domain they can also be effectively repelled by filtering them out in the same domain [6].

However, in the last few years several attacks have been investigated that, by mimicking the time-frequency power distribution of the legitimate signal can disrupt reception with a significantly lower transmit power.

Moreover, by adaptively steering the attack signal the threat can be made selective, in the sense that it only targets a specific receiver or a much smaller area. An example of this strategy was given in [7], where the jamming signal is designed to substantially cancel the legitimate signal while taking into account some degree of uncertainty in the measured pseudoranges.

10.2.4 Spoofing

Spoofing is a forging or illegitimate modification threat that can operate either at the physical or data link layer with different features.

Data message spoofing consists in transmitting a forged GNSS signal, where the data (e.g., ephemeris, ionospheric corrections) are chosen different from the authentic ones with the aim of luring the receiver into computing a false position, velocity or timing.

On the other hand spoofing at the physical layer aims at reproducing the signal from each single satellite in view with the exact same data as the legitimate one but with a different time of arrival at the victim receiver. This will affect the estimation of pseudoranges and hence the position computation.

Several distinct attacks in this class have been investigated in the literature, with the corresponding countermeasures: the simplest form of spoofing attack is called *meaconing*, where the attacker simply retransmits its overall received signal towards the victim, without the need for any processing, with a higher power than the legitimate one. The victim will acquire and lock on the attack signal, computing the attacker's position as its own, and this is the only position that can be induced on the victim, since the GNSS signals are retransmitted aligned exactly as they were received by the attacker.

In fact, in order to spoof a different, arbitrary position, it is necessary that the signals from satellites that are in view from the spoofed position are transmitted to the victim, aligned in the proper way. Since the parameters for constructing the open service GNSS signal are publicly available, this is in principle possible with a minimally sophisticated signal simulator.

The same task becomes more complex if the GNSS signal includes some cryptographic provision, in the form of secret or unpredictable code chips or data symbols. In this case, the attacker is required to observe its received signal, attempt to estimate the unknown symbols and replay the signal with the estimated symbols and the proper delay in what is called the *secret code estimation and replay (SCER)* attack [8, 9]. Observe from Figure 10.2 that some signals may need to be anticipated rather than delayed, if the corresponding satellite is closer to the spoofed position than to the attacker, making it apparently impossible to estimate the unknown symbol. The attacker can overcome this difficulty by introducing an additional common delay to all signals, resulting in a timing offset on the victim receiver that will not be detected, unless the receiver is already synchronized.

Moreover, in [10] the authors showed that it is typically possible to find a sufficient number of satellites in view for which the attack only requires

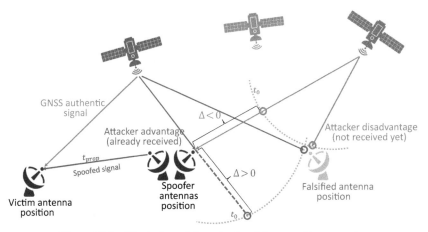

Figure 10.2   Illustration of the selective delay spoofing scenario.

delays, even with carefully synchronized receivers, and that anticipated estimation of the cryptographic symbols can leverage the redundancy provided by channel coding to some extent.

10.2.5 Self-attacks

A particular attention should be given to those cases where the attacker is colocated with the victim, as this will give the attacker the advantage of knowing the victim's channel in a very reliable manner and allow him to tailor the spoofed transmitted signal, e.g., to cancel the legitimate one.

For instance, GNSS based digital tachograph are envisioned to guarantee compliance by professional drivers to speed limits, road safety conduct, maximum driving hours per day. Hence, motivated by need to escape the regulatory constraints, or a close monitoring by his/her employer, a driver may aim at denying or even falsifying the position and velocity computation by the GNSS receiver.

Another self-attack context may come from the adoption of automatic road tolling procedures where the system on board the vehicle records passages through toll roads, so that the fees are automatically collected without the need for toll booths. There, systematically recording false vehicle trajectories may yield a medium-sized logistics company significant savings through toll evasion.

Similarly, the crew of a fishing vessel may have an interest in spoofing the GNSS signal towards their own receiver so that a false vessel trajectory is recorded by on board Vessel Monitoring Systems and, e.g., reported to a Fishing Monitoring Centre (FMC) as compliant to regulations, while the vessel instead is illegally fishing in closed areas.

For the above reason, it is not advisable to design integrity or availability protection mechanisms for GNSS under the assumption that the attacker is in a different position that the victim, and cannot observe the same channel. Similarly, one should not assume that the attack signal is severely attenuated before it reaches the victim receiver.

10.3 Navigation Message Authentication

One possible countermeasure to the spoofing of the navigation message at data level is to authenticate the content with a standardized mechanism for message authentication and integrity protection.

The broadcast nature of the GNSS message calls naturally for the use of asymmetric schemes (digital signature), where only the legitimate transmitter can sign the message, whereas any receiver can verify the message authenticity. Unfortunately, digital signature schemes require more complex verification, longer signatures and keys, such that it is unfeasible to have a signature transmitted (and verification performed) in each transmission packet.

On the other hand, the use of symmetric schemes, which would be more efficient both computationally and in terms of transmission resources, would bear along the complexity of designing a proper key management scheme. In fact, it is not possible to have a private symmetric key shared among all commercial receivers, as it would need to be stored in a tamper resistant module.

Hybrid solutions have therefore been sought, that could combine the best of both worlds. One such solution is the use of Timed Efficient Stream Loss Tolerant Authentication (TESLA), proposed in [11, 12] and borrowed from the Wireless Sensor Networks context, which resembles the GNSS one for its limited computation and transmission resources. TESLA is based on the use of a symmetric and efficient Message Authentication Code mechanism, in which the symmetric key is revealed in a broadcast manner after the MACs have been delivered to all receivers and immediately replaced by a new key, so that while any receiver can verify the message, it is too late for an attacker to use the disclosed key to forge past messages. Clearly this operating mode requires some degree of synchronization among transmitter and receivers,

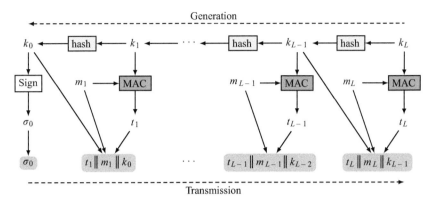

Figure 10.3   Illustration of the broadcast authentication scheme with delayed key release, based on TESLA (adapted from [14]). The key k_ℓ, that is used to compute the MAC t_ℓ from the message m_ℓ is transmitted later, together with the next message and code $(m_{\ell+1}, t_{\ell+1})$.

and bears the penalty of a delay in the message verification, which may be unacceptable for some applications.

Moreover, in order to authenticate each symmetric key k_ℓ, a commitment scheme is used where the commitment value for k_ℓ is transmitted and possibly authenticated together with the corresponding message and MAC (m_ℓ, t_ℓ). In the loss tolerant version of the protocol, the keys are computed backwards from k_L using a reverse one-way hash chain, that is $k_\ell = h(\ell, k_{\ell+1})$ where $h(\cdot)$ is a cryptographic hash function. Thus, each key k_ℓ can act as a commitment value for future keys $k_{\ell+1}, k_{\ell+2}, \ldots$, while the root key k_0 is authenticated through a separate digital signature scheme, that is therefore used only once in a while at the beginning of a new hash chain.

It should be noted, however, that using long hash chains leads to the piling up of collisions, and can linearly increase the success probability of collision-based attacks as was shown in [13].

Another possible solution, proposed in [14], is to make use of the notion of *signature amortization*, that is having a single signature authenticate several consecutive messages, and leveraging the fact that within the same IOD period, those messages are actually several copies of the same message.

By binding the messages together through a one-way hash chain, a short fraction of the signature can be transmitted with each packet. Here the length of each hash chain is limited by the IOD duration and hence even the collision accumulation effect is kept low.

It should be noted, however, that navigation message authentication cannot provide assurance that the received signal has not been illegitimately modified by altering its physical layer properties e.g., by artificially delaying it, replaying it at a slightly later time, or introducing an artificial Doppler shift. In fact, even if the transmitted navigaton data has a cryptographically generated, unpredictable part, the attacker can adopt a decode-and-replay scheme, as long as his carrier to noise ratio (C/N_0) alows for reliable decoding.

10.4 Cryptographic Integrity of the Navigation Signal

As we saw in the previous section, only a physical layer integrity protection mechanism can thwart attacks like de-alignment and selective delay of single GNSS signals.

In this section we will examine two possible solutions, one operating at the signal level, the other at the spreading code level.

10.4.1 Signal Watermarking

The technique of signal watermarking consists in introducing random modification to the transmitted signal, e.g., by changing some physical layer parameters (carrier frequency/phase, time delay/jitter) or by superimposing artificial noise, for limited time frames, so that data demodulation is not affected. Yet, after the introduced modification have been received, encoded in the authentic navigation message, the receiver can trace back the corresponding received signal and verify the presence of watermarks.

10.4.2 Partial Spreading Code Encryption

A possible solution to offer signal authentication with a delayed approach, was proposed in several distinct fashions in [15–18], and consists in partially encrypting the PRN spreading code $\mathbf{c} = [c_1, \ldots, c_L]$ and $c_i \in \{-1, 1\}$ of each SV. That is, to replace it with a new code $\mathbf{c}'$ which coincides with $\mathbf{c}$ for a large part, whereas it is unpredictably modified for a small number of the chips. As such, the peak value in the correlation between the local copy of $\mathbf{c}$ (that is known at the receiver) and the transmitted encrypted code $\mathbf{c}'$ will not exhibit a significant loss with respect to that in the autocorrelation of $\mathbf{c}$, and will still allow the receiver to acquire and track the signal, although a slightly higher C/N_0 will be necessary for lock-in.

On the other hand, the short encrypted part offers the possibility to subsequently verify the signal authenticity by means of a delayed release of the unpredictable information in the modified part.

- The modified code c' should be generated randomly according to a conditional distribution $p_{c'|c}(a|b)$. The amount of unpredictability of the modified code c' given the open public code c is measured by the *guessing entropy*

$$H_g(c'|c = b) = \log_{1/2} \max_a p_{c'|c}(a|b)$$

- The correlation loss ρ is measured as

$$\rho = 1 - \frac{\mathrm{E}\left[\sum_i c'_i c_i\right]}{\sum c_i^2} = \frac{2\,\mathrm{E}\left[d_{\mathrm{H}}(c', c)\right]}{L}$$

where $d_{\mathrm{H}}(c', c)$ represents the Hamming distance between the open and the (random) encrypted version of the code

- The amount of information that needs to be shared among the receivers to allow them to verify the integrity and authenticity of the signal is given by the *Hartley entropy*

$$H_0(c'|c = b) = \log_2 N$$

where N is the number of distinct possible realizations of c' given that $c = b$.

Clearly there is a trade-off to be chosen, in the design of such schemes. Minimizing the correlation loss and maximizing the encrpyted code unpredictability are conflicting objectives as they depend in an opposite way on the number of chips that remain equal in both codes. Moreover the Hartley entropy is never lower than the guessing entropy, with equality holding only for uniform vectors.

Without loss of optimality, the encrypted code can be generated as

$$c'_i = c_i x_i$$

where the random vector $x = [x_1, \ldots, x_L]$, with $x_i \in \{-1, 1\}$ is generated independently of c. Then, the above metrics can be written in terms of x as:

$$H_g(c'|c = b) = H_g(x) \tag{10.1}$$

$$\rho = \frac{2\,\mathrm{E}\left[y\right]}{L} \tag{10.2}$$

$$H_0(c'|c = b) = H_0(x) \tag{10.3}$$

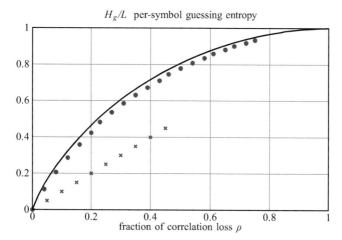

Figure 10.4 Illustration of the trade-off between guessing entropy and correlation loss for partial spreading code encryption. Our proposed solution is denoted by •, the proposals from [15–17] by × while the asymptotic upper bound as $L \to \infty$ is plotted in a solid line.

where y is itself a random variable expressing the Hamming weight (that is the number of chips set to -1) in x.

The optimal trade-off between the above requirements is then obtained by choosing some value of N and generating x as uniform among the N with lowest Hamming weight among the 2^L binary sequences of length L.

Such a trade-off is illustrated in Figure 10.4, where the asymptotic upper bound for $L \to \infty$ is drawn as a solid line, and also the results for several proposals in the literature are given.

10.5 Integrity Verification with Side Information

Mechanisms that pursue GNSS authentication and integrity protection can be categorized in two different approaches. The methods seen so far tackle the system architecture design and foresee a modification of the current state of the system, either in the signal-in-space (signal level integrity techniques), in the space segment or in the user segment (cryptographic techniques and receiver based techniques such as antenna arrays and dual frequency receivers). Other techniques exist that put aside GNSS design and rather exploit other independent sources of position redundancy to authenticate the final PVT. Most of the target applications for GNSS today integrate other useful capabilities such as movement sensors (IMU, tachograph), internet connection, telecommunication signal receivers (WiFi, GSM, UMTS, LTE, the so called signals of opportunity), cameras and so on. These additional features can complement GNSS receivers and provide alternative positioning information that can be exploited for security purposes. While the GNSS output is global position and velocity with respect to the inertial coordinate system, inertial sensors can provide information on the relative movement in relatively short time intervals. Similarly, cameras can be used to determine relative motion with techniques such as optical flow, Simultaneous Location and Mapping (SLAM) and visual odometry. These techniques provide information on the trajectory of the body with respect to a local reference frame rather than globally. Nonetheless they are harder to jam or spoof and tampering with all these diverse data exponentially increases the attacker's effort.

In [19] an assessment is performed on the current status of GNSS resilience to spoofing attacks in smartphones in order to verify the level of integration of this system with other positioning sources in mobile devices. In the specific case of smartphones the available sources for integrity cross-checks are various. The navigation message from the signal-in-space can be tested against authenticated data downloaded from the network (e.g., through Assisted GNSS). Cellular networks have intrinsic location capabilities and several standardized positioning methods exist that provide an authenticated estimate of the user's location. All smartphones today integrate inertial sensors and a magnetometer, allowing to estimate the change of the device pose over time and compare it with the GNSS output. The position history keeps track of the device displacement over time. It can be exploited in order to rule out trajectories that are physically impossible and thus mitigate basic position spoofing attacks. Several spoofing experiments were performed in [19] in a protected environment on multiple phones of several makes and models. The testing was performed through the synthesis of the GNSS signal of a different position. The data were selectively tampered with and completely wiped off, a moving trajectory was forced into the receiver while the device remained still and the system time was shifted forward or backward. Most of the testing was performed while the devices retained network connectivity.

The test campaign highlights that the available side information does not contribute to spoofing detection capabilities, as the application layer was never alerted of any anomaly when the phones were exposed to trivial spoofing attacks. The proposed integrity checks would thus enhance the protection of location based services, allowing the end user to safely exploit the emerging applications, that are currently exposed to the spoofing threat.

10.5.1 Binary Hypothesis Testing

The target of these integrity verification techniques is assessing whether the GNSS output is authentic or has been tampered with. Therefore the detection technique is required to output a binary indicator that states whether GNSS is subject to spoofing. This method is known as binary hypothesis testing and it is based on a Bayesian approach.

The aim of binary hypothesis testing in this specific scenario is to decide between the two hypotheses:

$\mathcal{H}_0$:   GNSS output is consistent with the actual trajectory

$\mathcal{H}_1$:   GNSS output has been tampered with

The decision algorithm takes as input all the M available positioning measurements and their joint probability density function conditioned on both hypotheses: $p(\mathbf{y_1}, \ldots, \mathbf{y_M}\vartheta|\mathcal{H}_0)$, $p(\mathbf{y_1}, \ldots, \mathbf{y_M}\vartheta|\mathcal{H}_1)$. In this formulation $\mathbf{y_i}$ denotes the measurement vector from source i while ϑ denotes the unknown quantities and parameters of the moving trajectory. The above probability density functions represent the likelihood of each hypothesis according to the measurements; following the Neyman-Pearson theory in [20], the decision algorithm is based on the ratio between the two likelihood quantities:

$$\mathcal{L}(\mathbf{y_1}, \ldots, \mathbf{y_M}, \vartheta) = \frac{p(\mathbf{y_1}, \ldots, \mathbf{y_M}, \vartheta|\mathcal{H}_0)}{p(\mathbf{y_1}, \ldots, \mathbf{y_M}, \vartheta|\mathcal{H}_1)}, \qquad (10.4)$$

and a threshold γ can be used for the final decision:

$$\hat{\mathcal{H}} = \begin{cases} \mathcal{H}_0 & \text{if} \quad \mathcal{L}(\mathbf{y_1}, \ldots, \mathbf{y_M}, \vartheta) > \gamma \\ \mathcal{H}_1 & \text{if} \quad \mathcal{L}(\mathbf{y_1}, \ldots, \mathbf{y_M}, \vartheta) < \gamma \end{cases}, \qquad (10.5)$$

As the parameters ϑ are in general not known, the probability density function can be replaced with the maximum likelihood estimate of ϑ,

$$\hat{\vartheta}_i = \arg \max p(\mathbf{y_1}, \ldots, \mathbf{y_M}|\vartheta, \mathcal{H}_i),$$

Moreover, as the measurement sources are often independent, the joint probability density function becomes:

$$p(\mathbf{y_1}, \ldots, \mathbf{y_M}|\vartheta, \mathcal{H}_i) = p(\mathbf{y_1}|\vartheta, \mathcal{H}_i)p(\mathbf{y_2}|\vartheta, \mathcal{H}_i) \ldots p(\mathbf{y_M}|\vartheta, \mathcal{H}_i). \quad (10.6)$$

10.5.2 Use of Inertial Measurement Systems

Spoofing detection techniques based on IMU measurements have been vastly used in safety critical applications such as aviation and maritime navigation, where GNSS spoofing is a serious threat. Several works have investigated the increase in security that high precision sensors can provide, [21, 22]. Some of the works focus on precise trajectory estimation by refining extended kalman filter algorithms and similar techniques (smoothing, particle filtering, etc.). Other rather focus on authentication techniques that exploit the randomization offered by unpredictable environment-related events, such as wind gusts and vehicle vibrations and oscillations. These high frequency phenomena provide random sequences that can be observed on inertial sensor measurements and looked for into GNSS measurements as a sort of environment-superimposed signature [21, 23]. This technique has been found to be beneficial both on high end IMUs [21] and on low cost ones, [24].

[24] targets an attacker with imperfect information on the vehicle's dynamics and develops a spoofing detection algorithm based on the decoupling of IMU and positioning obtained by GNSS, that allows a direct comparison of the acceleration for an unlimited time window. The proposed technique is reviewed in different application scenarios such as railways and automotive, wherein it is found to be less effective due to the lower intensity of high frequency components in the acceleration process.

A different approach is that of using IMU measurements to stir the GNSS tracking loop in order to make it more resilient to spoofing signals and more unlikely to lock on them [25].

The authors of [26] perform quality assessment of measurements taken with IMU and GNSS chipsets in different mobile phone models. They point out how even measurements from low-cost IMUs of mobile devices provide useful data not only for navigation integration but also for security enhancement, [24].

Spoofing detection in vehicular applications is investigated in [27], where a mobile device is used for comparing the absolute value of linear and angular accelerations with those obtained from the GNSS solution. This approach avoids the calibration of IMU and is invariant to manipulations to the device's initial orientation. The automotive scenario is also the target application in [28], wherein the proposed solution integrates data from GNSS, IMU, and odometer. Differently from [24] and [27], the comparison domain is position rather than acceleration, and the detection statistics are obtained as the norm of the difference between position vectors (from GNSS and from IMU/odometer). The novelty in this approach is the idea of performing GNSS based sensor calibration at fixed time intervals only if the spoofing detection algorithm confirms that the GNSS solution is authentic.

Some relevant implementation issues are pointed out in [24], such as data latency and the time synchronization between inertial sensor measurements and GNSS chipset data, which have in general different sampling frequencies. Moreover, a high enough sampling frequency is needed in order to capture all the motion ranges of the specific application, meaning that a higher sampling frequency is needed for high dynamics scenarios, as well as for authentication techniques based on high frequency components. As reported in [29] the computational complexity of the trajectory estimation algorithm based on IMUs is directly related to the non-linear nature of the orientation representation. In order to enhance the estimation performances, knowledge on the displacement process and other environment related constraints can be included in the estimation algorithm, leading to a better solution and possibly decreasing the computational burden (e.g., in automotive scenarios the degrees of freedom of the movement are limited with respect to drone scenarios).

10.5.3 Use of Visual Data

Another potential source of positioning information that is more and more ubiquitous as a complement to GNSS navigation is constituted by imagery and video from cameras mounted on the moving vehicle. Several works in the literature have focused on the integration of image processing techniques for enhancing navigation with GNSS, making it more resilient to multipath and degradation due to bad environment conditions (e.g., urban canyons). In [30] the authors develope a NLOS signal detector in urban environments by mapping the close-by obstacles detected with a fisheye camera mounted on the rooftop of a car.

As reported in [31], there are mainly two approaches for integrating visual information with GNSS positioning. The first consists in inferring the movement of the receiver from the camera images (i.e., dead reckoning), which can be performed through algorithms such as optical flow and Simultaneous Location and Mapping (SLAM).

The second approach consists in comparing the acquired imagery with a database of geo-referenced data in order to enhance the global position accuracy. An example is presented in [32], where sub-meter accuracy was obtained by performing pose estimation through comparison of the camera images with Google Street View geo-tagged 360° panoramas. The described algorithm is divided in two phases. The first consists in processing the database by extracting the feature representation of the images to efficiently perform queries. In the second phase the pose estimation is performed for each query image after extracting the best k matching images from the database.

In [33] the street View database is used for a similar purpose but with a different comparison mechanism. The feature extraction is performed for a whole video sequence, exploiting the temporal correlation between consecutive acquisitions. A probability distribution of the position is obtained and geo-localization is performed in a bayesian framework, exploiting a very simple constant-velocity motion model. The resulting estimate is not very precise due to the quantization effect of using geo referenced images sampled at roughly 12m distance, however the approach is very general and does not require any calibration or side information on the camera relative motion.

Another scenario in which integrated cameras can contribute to enhance security is drone applications. A study on the GNSS spoofing threats to UAVs was carried out in [34], where the feasibility of UAV capture was investigated through the simulation of several spoofing scenarios, highlighting the need for a reliable detection mechanism. The use of image processing algorithms for spoofing detection was discussed in [35], where the integration of the inertial measurement unit and the UAV camera was proposed in order to detect anomalies in the measured velocity. The velocities calculated through integration of the accelerometer measurements are fused with those obtained from the camera through the Lukas-Kanade algorithm, and the result is compared with the GNSS velocity measurements to decide whether the receiver is under spoofing.

The ideal anti-spoofing algorithm should allow to include in the detection strategy every available measurement related to the evolution of position in time either in the global or in the local coordinate frame. For each measurement vector, the noise model should be known in order to retrieve the probability density function of the data and apply the binary hypothesis test, that leads to the optimal solution.

References

[1] NovAtel, Inc. "Chapter 8. GNSS applications and equipment," *An introduction to* GNS, online at www.novatel.com/an-introduction-to-gnss/chapter-8-gnss-applications-and-equipment/applications/, as accessed on 25 March 2019.

[2] E. D. Kaplan and C. J. Hegarty, "Satellite Signal Acquisition, Tracking and Data Demodulation," *Understanding GPS: Principles and Applications*, Artech House Inc., 2006, pp. 153–241.

[3] H. Huang, G. Gartner, J. M. Krisp, M. Raubal and N. Van de Weghe, "Location based services: ongoing evolution and research agenda," *Journal of Location Based Services*, vol. 12, n. 2, pp. 63–93, 27 August 2018.

[4] K. C. Zeng, S. Liu, Y. Shu, D. Wang, H. Li, Y. Dou, G. Wang and Y. Yang, "All Your GPS Are Belong To Us: Towards Stealthy Manipulation of Road Navigation Systems," *27th USENIX Security Symposium*, Baltimore, MD, 15–17 August 2018, pp. 1527–1544.

[5] R. Abbas, K. Michael, and M. G. Michael, "The Regulatory Considerations and Ethical Dilemmas of Location-Based Services (LBS): A Literature Review," *Information Technology & People* 27 (1): 220. doi:10.1108/ITP-12-2012-0156, 2012.

[6] D. Borio and P. Closas, "A Fresh Look at GNSS Anti-Jamming", *Inside GNSS*, vol. 12, n. 5, pp. 54–61, September 2017.

[7] G. Caparra, S. Ceccato, F. Formaggio, N. Laurenti, and S. Tomasin, "Low Power Selective Denial of Service Attacks Against GNSS," *International Technical Meeting of the Satellite Division of The Institute of Navigation, ION GNSS+ 2018*, 2018, pp. 3028–3041.

[8] T. E. Humphreys, "Detection Strategy for Cryptographic GNSS Anti-Spoofing," *IEEE Transactions on Aerospace and Electronic Systems*, vol. 49, n. 2, pp. 1073–1090, April 2013.

[9] G. Caparra, N. Laurenti, R. T. Ioannides, and M. Crisci, "Improving Secure Code Estimation and Replay Attack and Detection on GNSS Signals," *ESA Workshop on Satellite Navigation Technologies and European Workshop on GNSS Signals and Signal Processing, NAVITEC*, 2014.

[10] G. Caparra, S. Ceccato, N. Laurenti, and J. Cramer, "Feasibility and Limitations of Self-Spoofing Attacks on GNSS Signals with Message

Authentication," *International Technical Meeting of The Satellite Division of the Institute of Navigation, ION GNSS+ 2017*, Portland, OR, 25–29 September 2017, pp. 3968–3984.

[11] P. Walker et al., "Galileo Open Service Authentication: A Complete Service Design and Provision Analysis," *Proceedings of the 28th International Technical Meeting of the Satellite Division of the Institute of Navigation (ION GNSS+ 2015)*, 2015.

[12] I. Fernández-Hernández, G. Seco-Granados, I. Rodríguez, and J. D. Calle, "A Navigation Message Authentication Proposal for the Galileo Open Service," *NAVIGATION: Journal of The Institute of Navigation*, vol. 63, n. 1, pp. 85–102, January 2016.

[13] G. Caparra, S. Sturaro, N. Laurenti, and C. Wullems, "Evaluating the security of one-way key chains in TESLA-based GNSS Navigation Message Authentication schemes," *2016 International Conference on Localization and GNSS, ICL-GNSS 2016*, 2016, pp. 1–6.

[14] G. Caparra, S. Sturaro, N. Laurenti, C.Wullems, and R. T. Ioannides, "A Novel Navigation Message Authentication Scheme for GNSS Open Service," *International Technical Meeting of The Satellite Division of the Institute of Navigation, ION GNSS+ 2016*, Portland, OR, 12–16 September 2016, pp. 2938–2947.

[15] L. Scott, "Anti-spoofing and authenticated signal architectures for civil navigation systems," *Proceedings of the Institute of Navigation GPS/GNSS 2003 conference*, 2003, pp. 1543–1552.

[16] O. Pozzobon, L. Canzian, M. Danieletto, and A. Dalla Chiara, "Anti-spoofing and open GNSS signal authentication with signal authentication sequences," *ESA Workshop on Satellite Navigation Technologies and European Workshop on GNSS Signals and Signal Processing, NAVITEC, 2010*, pp. 1–6.

[17] J. M. Anderson et al., "Chips-Message Robust Authentication (Chimera) for GPS Civilian Signals," *International Technical Meeting of The Satellite Division of the Institute of Navigation, ION GNSS+ 2017*, Portland, OR, 25–29 September 2017, pp. 2388–2416.

[18] B. Motella, D. Margaria, and M. Paonni, "SNAP: An authentication concept for the Galileo open service," *2018 IEEE/ION Position, Location and Navigation Symposium (PLANS)*, 2018, pp. 967–977.

[19] S. Ceccato, F. Formaggio, G. Caparra, N. Laurenti, and S. Tomasin, "Exploiting side-information for resilient GNSS positioning in mobile phones," *2018 IEEE/ION Position, Location and Navigation Symposium, PLANS 2018*, Monterey, CA, 23–26 April 2018, pp. 1515–1524.

[20] S. M. Kay, *Fundamentals of statistical signal processing, Vol. II: Detection Theory, Signal Processing*. Upper Saddle River, NJ: Prentice Hall, 1998.

[21] C. Tanil, S. Khanafseh and B. Pervan, "Impact of Wind Gusts on Detectability of GPS Spoofing Attacks Using RAIM with INS Coupling," *Proceedings of the ION 2015 Pacific PNT Meeting*, 2015.

[22] C. Tanil, S. Khanafseh and B. Pervan, "An INS Monitor against GNSS Spoofing Attacks during GBAS and SBAS-assisted Aircraft Landing Approaches," *Proceedings of the 29th International Technical Meeting of the Satellite Division of The Institute of Navigation (ION GNSS+ 2016)*, 2016.

[23] P. F. Swaszek, S. A. Pratz, B. N. Arocho, K. C. Seals, R. J. Hartnett, "GNSS Spoof Detection Using Shipboard IMU Measurements," *Proceedings of the 27th International Technical Meeting of The Satellite Division of the Institute of Navigation (ION GNSS+ 2014)*, 2014.

[24] S. Lo, Y. H. Chen, T. Reid, A. Perkins, T. Walter and P. Enge, "Keynote: The Benefits of Low Cost Accelerometers for GNSS Anti-Spoofing," *Proceedings of the ION 2017 Pacific PNT Meeting*, 2017.

[25] G. Falco, M. Pini and G. Marucco, "Loose and Tight GNSS/INS Integrations: Comparison of Performance Assessed in Real Urban Scenario," *Sensors*, 2017.

[26] V. Gikas and H. Perakis, "Keynote: The Benefits of Low Cost Accelerometers for GNSS Anti-Spoofing," *Proceedings of the ION 2017 Pacific PNT Meeting*, 2017.

[27] J. T. Curran and A. Broumandan, "On the use of Low-Cost IMUs for GNSS Spoofing Detection in Vehicular Applications," *International Technical Symposium on Navigation and Timing (ITSNT)*, 2017.

[28] A. Broumandan and G. Lachapelle, "Spoofing Detection Using GNSS/INS/Odometer Coupling for Vehicular Navigation," *Sensors*, 2018.

[29] M. Kok, J. D. Hol and T. B. Schön, "Using Inertial Sensors for Position and Orientation Estimation," *Foundations and Trends in Signal Processing*, 2017.

[30] J. Marais, S. Ambellouis, C. Meurie, J. Moreau, A. Flancquart, and Y. Ruiche, "Image processing for a more accurate gnss-based positioning in urban environment," *22nd ITS World Congress*, 2015.

[31] P. D. Groves, "The PNT Boom. Future trends in Integrated Navigation," *InsideGNSS*, 2013.

[32] L. Yu, C. Joly, G. Bresson, and F. Moutarde, "Monocular urban localization using street view," *2016 14th International Conference on Control, Automation, Robotics and Vision (ICARCV)*, 2016.

[33] G. V. Castano, A. R. Zamir and M. Shah, "City scale geo-spatial trajectory estimation of a moving camera," *2012 IEEE Conference on Computer Vision and Pattern Recognition*, 2012.

[34] A. J. Kerns, D. P. Shepard, J. A. Bhatti, and T. E. Humphreys, "Unmanned aircraft capture and control via GPS spoofing," *Journal of Field Robotics*, vol. 31, pp. 617–636, 2014.

[35] Y. Qiao, Y. Zhang, and X. Du, "A vision-based GPS-spoofing detection method for small UAVs," *13th International Conference on Computational Intelligence and Security (CIS)*, pp. 312–316, 2017.

11

CONASENSE Towards Future Technologies for Business Ecosystem Innovation

Peter Lindgren

Aarhus University, Denmark
E-mail: peterli@btech.au.dk

"Future Technologies for Multi Business model Ecosystem"
Where are the business models?

Future Integration of Communications, Navigation and Sensing technologies (CONASENSE) are challenges by finding profitable business models (BM). 5G technologies are very expensive to develop and implement and the prices to buy 5G licenses are high and expected to become even higher. On top of this classical BM's – e.g. the subscription BM – will not/maybe never be able to give enough turnover and profit – because cost of business model innovation and implementing 5G will be extremely high. Return on investment (ROI) will within required time span based on subscription BM for tele operators and other 5G investors not be possible to be achieve by mobile operators BM's alone [1, 2]. The challenge of 5G BM's are present for existing large and emerging Business Model Ecosystems (BMES) – like smart cities – but it is even more actual for rural areas where no or very poor technology – especially CONASENSE technology infrastructure exists. 5G technologies will therefore face difficulties to be implemented in rural areas – also in western countries – if businesses and their allies do not begin to think out of the box – and begin to adapt the approach of multi business modelling.

Business operating and who intend to operate in these future CONASENSE technology BMES seems however and although still to invest heavily in 5G standards, patents, research, technologies and even to some extent also – infrastructure. Several speakers at the WWRF conference in October 2018 in Denmark – wwrf.ch – documented this very clearly. Many

Security within CONASENSE Paragon, 167–184.

speakers also addressed – why this takes place – and even why some operators plan to do this even in the most rural areas. Why not offer rural areas less advanced technologies like 3G or 3,5 G like the business Blue Town does – www.Bluetown.com

5G technology is today under early testing and poised to be launched heavily soon. Mobile operators are preparing with a mixture of resignation and anticipation and they seem to know that this revolutionary technology will open new opportunities – BM's – capturing value from new 5G use cases and a widespread of adoption of the Internet of Things (IoT).

At the same time, the mobile operators are aware that they'll have to increase their existing technology infrastructure investments even more. They still have to upgrade their 4G networks to cope with radical growing communication demand. The future wireless network (FWN) will need to efficiently and flexibly provide diversified services such as: enhanced mobile broadband, ultra-reliable, low-latency communications and massive machine type communications. It should support a framework of multiple operational standards, coordinate a heterogeneous network with different types of base stations (BSs), process information generating from a huge volume of traffic, stay robust against all potential security threats, and support intelligent informed decisions by adapting to appropriate network functionality under constraints of time-varying workload and diverse user devices to meet guarantees.

In an analysis from Mackinsey [1] in 2018 prediction was made that network-related capital expenditures would have to increase by 60 percent from 2020 through 2025, doubling total cost of ownership (TCO) during that period. How can this happen? – What are the BMs? and have the Mobile operators lost their minds? Do the Mobile operators not see profit any longer as an important output of BM? – or Are we – standing outside the Future CONASENSE Technology BMES – just not capable to "see" the BM's of tomorrow and "see" the BM's that the mobile operators "see"?

This conundrum raises important doubt and questions about possible future BM's of CONASENSE technologies, future business model innovation (BMI) in future BMES related to CONASENSE technologies, business and investment strategies together with future value formulas of businesses BM's related to CONASENSE technologies.

In this book chapter, we focus on unwrapping future CONASENSE BM's and BMES's required to "carry" the future CONASENSE technologies. The chapter also discuss how BMI in a world of future CONASENSE technologies and related BMES's is expected to be carried out and implemented.

11.1 Introduction

Even if mobile operators delay heavy CONASENSE 5G investments, they will still need to increase infrastructure spending to cope with futures growing traffic on old technology platforms as 3G and 4G. There is no reason to believe that the historical increase of 20 to 50 percent traffic per year on existing network structure will change – or not happen. However, they could even fall on the higher end of that scale, which would affect investment intensity tremendously.

In an analysis of one European country, where all three operators followed a conservative approach to CONANSENSE 5G investment, it was predicted that total cost of ownership for Radio Access Network (RAN) – core technologies, that connects devices to networks and is a major part of modern 3G and 4G telecommunications – would increase significantly in the period from 2020 through 2025. Compared to the expected 2018 level as shown in Figure 11.1 the increase in cost would be very large.

For instance, in a scenario that assumes 25 percent annual data growth, TCO would expectedly rise by about 60 percent. In Figure 11.1, McKinsey estimated how cost will evolve if data growth 25%, 35% and 50%.

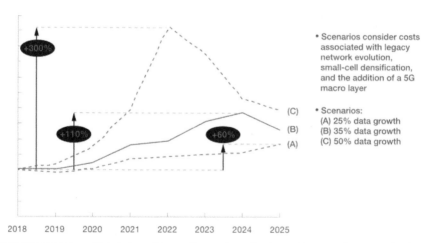

Total cost of ownership for mobile access networks will increase.

McKinsey&Company

Figure 11.1 Estimate of total cost of ownership (TCO) for mobile access network Mckinsey company 2018 [1].

Turnover at same time will expectably not follow the same evolvement – which means critical to "bleeding" business – to operators in general. Especially with a conservative look to BMI and business strategy the future 5G technology looks very risky.

Network sharing and CONASENSE 5G technologies [2] could maybe reduce cost and risk but still the future BM's, value formula's and business cases have not been shown clearly in any business case. As seen in table 2 the numbers of active network-sharing agreements have increased dramatically since 2010.

Businesses in the CONASENSE BMES taking the opportunity of network based business models (NBBM) [3] are increasing with exponential speed their BM's (Figure 11.2) and are coursing major disruption of businesses, BM BMESs [4–6] and even societies. The new NBBMs worries many governments and business managers responsible for BMI because these BM's are

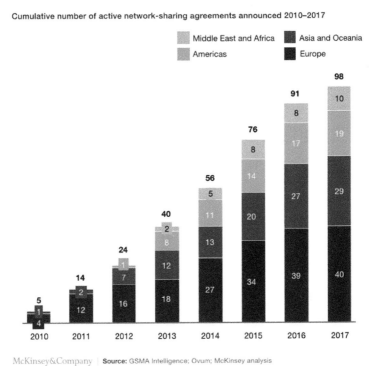

Cumulative number of active network-sharing agreements announced 2010–2017

McKinsey&Company   **Source:** GSMA Intelligence; Ovum; McKinsey analysis

Figure 11.2 Cumulative number of active network-sharing agreements announced 2010–2017. Mckinsey company 2018 [2].

1. Offering BMES and their related customers BM's for other value than money.
2. Are playing "the multi business model approach game" [7] – offering some BM for very low prices and earning their revenue on other BM either openly or hidden.
3. Are already merging themselves into various aspects of digital connected life.

This makes it very difficult for established businesses to analyze the BMES and compete with traditional profit oriented single based BM's approach and value formula's.

Cloud services BMES and information BMES are becoming influential and critical in routine sustainability of businesses. Low operational costs of IoT and Industry 4.0 driven future BMES will redefine the global economics setting the cornerstone of future BMs and Multi Business Model Innovation (MBMI).

The previous BMs are largely dependent upon traditional service such as messaging, voice and data. Future CONASENSE 5G and beyond technology will bring a wave of BMES changes due to massive digitalization of services. This will require novel BMs and new BMES regulation. These massive numbers of connected devices as seen in Figure 11.3 will be able

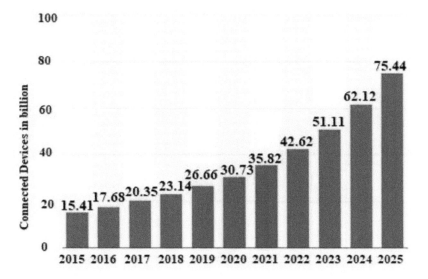

Figure 11.3 Estimated connected devices in billion 2015–2025 Alam Tanweer.

to connect with one another forming a device-to-device (D2D), device-to-machine (D2M) or machine-to-machine (M2M) network. At present there are not much research on the typologies and volumes of BM's for advanced dynamic NBBM between D2D, D2M, M2M and D2H.

Many of the new BM's are very difficult to understand, track and trace. These often cannot be understood and innovated with previous BM tools and BMI approach. We propose instead to adapt a Multi Business Model Innovation (MBMI) approach combined with a cross-interdisciplinary approach–like the Conasence approach.

11.2 Vertical and Horizontal BMES with Future CONASENSE

Investigating the NBBM deeper address that these new BM's are not just focused on vertical BMES's as seen in Figure 11.4 but also focus and are

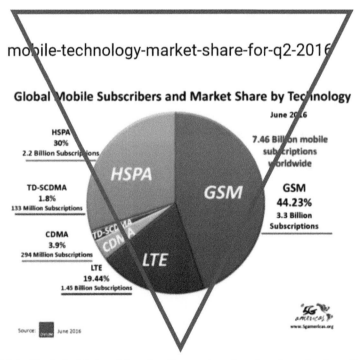

Figure 11.4   Vertical global mobile subscribers and market share by technology BMES q2 2016 adapted from [5].

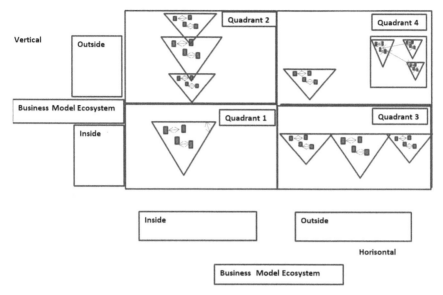

Figure 11.5 Vertical, horizontal and diversified BMES.

capable of offering itself to Horizontal BMES as seen in Figure 11.5 and even BMES that are diversified to focal BMES to the mobile operators as also shown in Figure 11.5.

Classically business – in this case mobile operators in vertical BMES Figure 11.4 – would do BMI for the vertical BMES Figure 11.5 quarant 1 and 2 – as e.g. Mobile operators try to innovate 5G conansense technologies for the mobile subscription BMES as shown in Figure 11.4. Following are highlights from the first quarter 2018 edition of the IHS Markit Mobile Infrastructure Market Tracker, which includes data for quarter ended December 31, 2017. The report tracks 2G, 3G, LTE and 5G equipment and software. Globally, the mobile infrastructure BMES – including all types of radio access network (RAN), switching and core equipment – which grew 15 percent sequentially in the fourth quarter of 2017 (Q4 2017), driven by previous year-end loaded projects in various spots, including China and India, and the completion of other expansion projects in North America. However as can be seen, the Q4 2017 market was down 10 percent on a year-over-year basis and down 14 percent for the full-year 2017. The worldwide mobile infrastructure BMES gained some steam in the fourth quarter, but it was not enough to bring the whole year back into positive territory [6]. The quarterly mobile infrastructure report tracks more than 50 categories of equipment, software

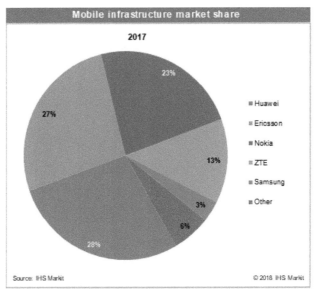

Figure 11.6   Mobile infrastructure market highlights.

and subscribers based on all existing generations of wireless network technology, including radio access networks, base transceiver stations, mobile soft switching, packet core equipment and E-UTRAN macro cells. It provides worldwide and regional BMES size, vendor BMES share, forecasts through 2022, in-depth analysis and trends.

Despite relative stability in North America and a slight pickup in Japan, full-year 2017 vertical mobile infrastructure BMES revenue declined in all regions, and even more so in China. All of this confirms that the vertical mobile infrastructure BMES are now red ocean, in the second year of the post-LTE peak era and are under extremely high pressure on turnover, earning and now also cost to to necessary investments as can be seen in small slices of the vertical mobile infrastructure BMES beneath.

- For the full-year 2017, global 2G/3G/LTE macro hardware revenue totaled $37.2 billion, declining 14 percent from $43.3 billion in 2016
- China remains the largest mobile infrastructure BMES with around 680,000 enhanced NodeBs (eNBs) shipped last year, down from 1 million in 2016
- When excluding 5G from the forecast, the global 2G/3G/LTE mobile infrastructure hardware BMES is projected to slowly decline to

$13 billion by 2022 – a five-year (2018–2022) compound annual growth rate (CAGR) of −18.4 percent
- When including 5G in the forecast, 2G/3G/4G/5G hardware revenue is anticipated to increase to $25 billion by 2022, with a five-year CAGR of −7.8 percent
- 5G hardware revenue is expected to reach $11 billion in 2022, starting from a very low base of early adopters in the US in the second half of 2018, followed in 2019 by South Korea and China's massive trial that will generate revenue for vendors.

Both positive and negative effects can be registered in the vertical BMES – but overall and seen as the vertical BMES (Figure 11.4) the BMES can be classified as "Red Ocean BMES" [8]. The impact of Conasence 5G to existing BMES if highly needed to support the mobile operators BMES and potentially get them out of the Red Ocean. Conasence concern in this business modelling sense is however

> Do the mobile operators actually get better business and business models in the vertical BMES? – or do they have to change their BM, BMI and BMES mindset and strategy?

Businesses in the Mobile operator BMES has for many years classically put most emphasis on classical BM like subscription BM in vertical BMES. The fast development of more Conansense 5G BMs like "sensoring" and "persuasive" BMs increasingly run autonomously by machines or technologies, these businesses will have to change business – and innovate and operate their BM's into new business model and new BMES – like horizontal BMES quadrant 3 and Diversified BMES quadrant 4 as seen in Figure 11.5 to survive.

CONASENSE 5G Technology will enable businesses to innovate BMs and thereby their businesses into businesses and BMES that are more embedded with AI and run virtually – and run in to BMES that have not original and strategically been their business, intended to be their business and definitely not known to the mobile operators before. Business Model Innovation Leadership (BMIL) therefore becomes a very important topic in a CONASENSE perspective – in other words

> Who should lead BMIL in a world to 5G CONASENSE technology and BM's? – humans, machines or both together.

> Will it be possible in the future even to carry out BMIL in future Business and BMI with 5G CONASENSE Technology.

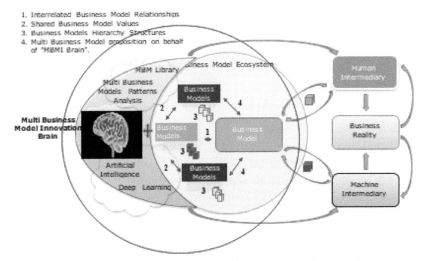

Multi Business Model Innovation Brain interaction supported by AI, Deep Learning, Multi Business Model Patterns Analysis and Multi Business Model Library

1. Interrelated Business Model Relationships
2. Shared Business Model Values
3. Business Models Hierarchy Structures
4. Multi Business Model proposition on behalf of "MBMI Brain".

Figure 11.7   Future CONASENSE 5G business modelling setup [11].

As BMs becomes more and more virtual the businesses classical bonds to BMs will be supplemented and even also exchanged with new bonds based on machine to BM's communication and machine to machine BM communication [18, 19]. This change will as a consequence have disruptive impact on the BM's between humans, humans and machines and machines to machines BM's. Also as we expects it will over the years influence the original business managers ability to "see" and "sense" BM's – in this case Managers of Mobile Operators.

Some "seeing" and "sensing" BM managing practice and abilities are expected to vanish over the years – and new will have to be learned and adapted – however in different ways. Gross Interdiciplinarity – the focus of CTIF GLOBAL CAPSULE (CGC) [10] and experimental design will be important ways to follow up to businesses and their existing and upcoming BM's and BMES. In Figure 11.7 we propose an experimental design of future CONASENSE 5G Business Modelling setup.

Today businesses already see the first results of this evolvement – strong business dependency of mobile telephone, sensors, social digital networks and an increasing trend of businesses solely communicating with technology

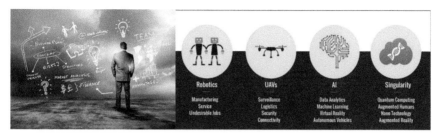

Figure 11.8 Previous business practice and Figure 3.7 future business practice with CONASENSE 5G business modelling setup [12].

and not or very little with humans in businesses and physical businesses. More and more businesses do not meet businesses physically and this calls businesses to change their BM's to meet users and customers requirements in virtual BMES space.

This change and evolvement in business practice influence businesses ability to "see", "sense", "relate" and "communicate" with other businesses.

It also questions highly reality in future of our business life and reality of our business practice – measured with our traditional and classical measurements of business practices based primarily on profit and low costs. Is it always optimal to focus on profit and lowest possible cost as many businesses do today – even our public and societal businesses? – or should we adapt new ways and practices to business effectiveness and efficiency.

The above mentioned also influence and change businesses way to do business – as e.g. more businesses will be dependent on advanced technologies. In this case 5G CONASENSE technology might in the future be the fundamental platform to do business Many public businesses – Municipalities and Ministries – together with private businesses – Banks, Insurance Businesses – can or will not accept cash in near future. Still they are not ready to adapt to crypto currency – but some kinds of "digital money" has to be accepted to move business into the new 5G conansense technology world. Previous currency are simply to slow, inconvenient and will not be able to match the requirements of 5G technology based business.

Today several businesses do only accept business communication and interaction as digital communication or in the future as robot to business communication. This development will continue with exponential speed and change business bonds to our businesses, humans and our society?

The exponential development of artificial intelligence technologies and persuasive technologies [13] enables persuasive BMs [14] in all kinds of BMI and push even more to new practice of operating and innovating BMs.

Picture 11.1   Future business with robots.

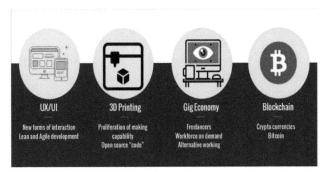

Picture 11.2   Future business with robots [12].

Regulators and businesses responsible of datasecurity and business etics struggles to meet and coop up with this exponential development in the favour of "protecting" the human – but at the same time without hurting businesses and preventing businesses to use the new potentials. Growth and workplaces it still important measurements to our society. As physical, digital and virtual worlds integrates – "melts together" in new Business Model Ecosystems (BMES) businesses will be able to interact with product-, service-, productions- and process technologies [14] independent of place, thing, body and time. Businesses "core business" will then be questioned continuously and classical businesses as e.g. a bank or an electricity business – even a hospital – will change or have to change "core business" more often. Value formula based on money will be transferred to other value formula based on other values. Classical services as subscriptions BM's will be based on business models for free and money will have to be earned on other business models.

Data driven business models [15] could be another new era in "5G CONASENSE Business Modelling" and in this case deep knowledge on communication, navigation, sensing technologies and advanced business model will play an important role enabling these BM's to be implemented to the BMES. Data driven BM's will potentially could transform businesses, reduce businesses risks, and create advantage for those businesses who knows how to use data effectively and efficiently – or have access to businesses that provide this big data services. As more businesses find ways to use the vast amounts of data they and their related BM's collect every day – they will indeed gain a competitive edge – but only if they know how to handle and use the data. Using data as the primary decision-making tool will make sense and will not just be "a trending term" as it is often seen today. Businesses will dig deep into data and form their business models on behalf of these. Traditional business executives will be challenged by making decisions based on not just gut instincts and narrow observations – but on big data analysis [16, 20]. Herein lays a great business service opportunity for Mobile Operators.

These new BM'S the mobile operators will have to find in a future of 5G technology BMES – but as seen above the potential is large especially when mobile operators begin to combine their subscription BM's with other BM's in all types of BMES – the multi business model approach.

It will also make humans and technologies with different competences better communicate, navigate and sense. Table 11.1 indicates beneath some of the benefits.

5G CONASENSE MBMI will in future be carried out with a mix of human and machine interaction – and machines will take over more and more MBMI – and not just routine MBMI. Mobile operators that do not realize in time or have not accepted that MBMI is and will be carried out differently to previous accepted MBMI forms will vanish or be in risk of being disrupted. Some would say that future 5G MBMI will be quiet different, even maybe unfair, not social and business sustainable to those previous MBMI forms and types known. This is however a fact when Mobile operators do not see and sense the 5G CONASENSE BM's provided by the 5 G technologies in a different way and with a different BM's mindset. Many Mobile Operators fear this new evolution, because they can not or do not "see" and "sense" the "counterpart" – but with a new BM's mindset they will learn how to "capitalize" on the potential of this huge MBMI potential embedded and support with these new 5G technologies?

Businesses "seeing", "sensing" and actually realizing MBMI with existing and from the past MBMI approaches will however not be capable

Table 11.1   Overall benefit categories of 5G conansense technologies and business model approach – adapted from Lindgren 2017 [7]

Over all benefit categories of future common BM language related to 5G Conansence technology business model approach	Over all benefits described in details.
Interoperability in MBMI	– ability of 5G CONASENSE technologies and BMs to work and innovate together relied on BMs complying with standard language of BM [17]
Support of government policies and legislation in MBMI	– ability to have standards, IPR and Patents of BMs to play a new central role in the global and regional 5G CONASENSE technologies BMES policy. Standards, IPR and Patents will frequently be referenced by regulators and legislators for protecting user and business interests, and to support government policies
Increase in interdisciplinary Business Modelling across vertical and horizontal BMES	– ability to increase interdisciplinary Business Modelling across vertical and horizontal physical, digital, virtual BMES with 5G CONASENSE Technology due to better possibility to "communicate", "navigate", "see" and "sense" together
Increase in 5G CONASENSE MBMI Technology development	– ability to provide a solid foundation for innovating new 5G MBMI technologies, new learning and new knowledge on BM and BMI to enhance and advance existing BMI practices
Provide economies of scale in MBMI projects	– ability to provide business being able to "produce" and "innovate" "large bats" and invest in "mass production" of BM's
Encourage businesses and society to do more and better MBMI	– ability to provide business develop MBMI on behalf of standards of sustainability, etics and humanity
Increase awareness of technical developments and initiatives within MBMI and 5G CONASENSE MBMI technologies	– ability to provide new and better technology based MBMI platforms hereby increasing awareness and providing a larger variety of accessible BMs to users, customers and businesses in general

(Continued)

Table 11.1 (Continued)

User, Consumer, network and "things" choice of BM and BMI would be easier to adapt	– ability to provide the foundation for new features and options, thus contributing to the enhancement of daily MBMI, user driven MBMI, interdisciplinary MBMI with both humans and things
Safety, reliability and etics in 5G MBMI technologies	– ability to ensure safety, reliability, etics and business care. As a result make users, customers, network, competences and businesses in general perceive standardized BM language as more valuable, dependable –raise these stakeholders confidence and take-up of new technologies and business models
Advance 5G MBMI Technologies	– ability to provide a solid foundation for research, learning and new knowledge on BM and BMI to enhance and advance existing BMI practices.

for operating in the described future MBMI BMES. Businesses that are not advanced enough – or do not have a MBMI framework model to support them and capture the potential of the new MBMI approach, the new 5G CONASENSE MBMI technologies will get difficulties. 5G CONASENSE business model approach will point to some of the solutions to deal with new types of communication, navigation and sensing technologies and business models in a world of 5G CONASENSE technologies.

References

[1] Mckinsey report 2018. https://www.mckinsey.com/industries/telecommu nications/our-insights/the-road-to-5g-the-inevitable-growth-of-infrastru cture-cost

[2] Mckinsey report 2018. https://www.mckinsey.com/industries/telecommu nications/our-insights/network-sharing-and-5g-a-turning-point-for-lone-riders

[3] Lindgren P, Yariv Taran and Harry Boer (2010). From single firm to network-based business model innovation International Journal of Entrepreneurship and Innovation Management 12(2) DOI: 10.1504/IJEIM.2010.034417.

[4] Lindgren, P and Ole Horn Rasmussen (2016). The Business Model Ecosystem Journal of Multi BMI, River Publisher.

[5] https://telecomlead-dwvighbb5mwqnytmb.netdna-ssl.com/wp-content/uploads/2016/09/Mobile-technology-market-share-for-Q2-2016.jpg

[6] Téral Stéphane (2018). Mobile infrastructure and carrier economics research at IHS Markit – https://technology.ihs.com/600864/global-mobile-infrastructure-market-down-14-percent-from-a-year-ago.

[7] Lindgren Peter (2017). Advanced Business Model Innovation Journal of Wireless Communication Springer.

[8] Tanweer Alam (2018). A Reliable Communication Framework and Its Use in Internet of Things (IoT).

[9] Kim, W.C. and Mauborgne, R. (2017). Blue Ocean Strategy with Harvard Business Review Classic Articles "Blue Ocean Leadership" and "Red Ocean Traps" (3 Books). Boston: Harvard Business School Press.

[10] CTIF GLOBAL CAPSUL – https://www.ctifglobalcapsule.org/about-cgc/

[11] Lindgren P., Per Valter and Ramjee Prasad (2019). "Digitizing Human Behavior with wireless sensors in Biogas 2020 Technological Business Model Innovation challenges" Journal of Personal Wireless Communication Springer Verlag.

[12] Perry Timms (2016). Slides from opening keynote to IPM South Africa – 2016 Annual Convention Perry Timms, Founder and Chief Energy Officer: PTHR (People and Transformational HR Ltd) Published in: Business License: CC Attribution-NonCommercial-ShareAlike License.

[13] Fogg, B. J. (2003). Persuasive technology: Using computers to change what we think and do. San Francisco, CA, USA: Morgan Kaufmann Publishers.

[14] Lindgren P., and Katharina Wuropulos (2017). Secure Persuasive Business Models and Business Model Innovation in a World of 5G Journal of Wireless Personal Communication October 2017, Volume 96, Issue 3, pp. 3569–3583|Springer Verlag.

[15] Hartmann, Philipp Max, Mohamed Zaki, Niels Feldmann and Andy Neely (2014). Big Data for Big Business? A Taxonomy of Data-driven Business Models used by Start-up Firms Cambridge Service Alliance: Linkedin Group: Cambridge Service Alliance www.cambridgeservicealliance.org

[16] Philipp Max Hartmann, Mohamed Zaki, Niels Feldmann, Andy Neely, (2016). Capturing value from big data – a taxonomy of data-driven business models used by start-up firms International Journal of Operations and Production Management Volume 36, Issue 10.

[17] Lindgren P (2018). The Multi Business Model Approach Part 1, River Publishers.

[18] Snapp, Jeffrey (2017). The Library Beyond the Book cognition, Digital_ Humanities, knowledge design, libraries, library design, networked culture, the future of the book – http://jeffreyschnapp.com/2013/09/28/the-library-beyond-the-book/

[19] Snapp, Jeffrey (2017). http://www.gsd.harvard.edu/course/what-should-or- could-scholarly-knowledge-look-like-in-the-21st-century-a-knowledge-design-seminar-spring- 2017/

[20] O. Acker, et al. (2013),"Benefiting From Big Data: A New Approach for the Telecom Industry".

Index

About the Authors

Albena Mihovska, obtained her PhD from Aalborg University (2008), and is an Associate Professor at Aarhus University, Department of Business Development and Technologies, where she is with the CTIF Global Capsule research group. Her main current activities relate to research in the area of smart dense connectivity and related applications; 5G ultradense access networks, Internet of Things technologies for healthcare and smart grid, and most recently, holographic communications. In addition, she focuses on aspects of digital innovations and their impact on business. Albena Mihovska has active teaching and supervision activities of Bachelor, Master and PhD students as well as strong participation and leadership of EU-funded projects. She has been an active contributor to ITU-T and ETSI standardiaztion activities in the areas of Internet of Things. She is a member of IEEE, and INFORMS (the leading international association for Operations Research & Analytics professionals). She has authored and co-authored more than 170 publications, including peer-reviewed international books, journal and conferences publications.

Dimitriya Mihaylova received her BSc and MSc degrees in Telecommunications from the Technical University of Sofia (TUS), Bulgaria, in 2014 and 2016 respectively. From May to August 2014 she took part in the Erasmus educational program in the Universidad de Las Palmas de Gran Canaria, Spain, where she completed her Bachelor's thesis on the topic of "Development of a communication system for ArduCopter model through the Iridium Satellite Network". From September 2015 to February 2016, Ms Mihaylova developed her Master's thesis project titled "Evaluation of the Dielectric Properties of Materials at High Temperatures" in the Universitat Politècnica de València, Spain, her second Erasmus participation. Since 2016, she has been with the Communication Networks Department of the Technical University of Sofia, where she is an Assistant Professor and a PhD student. Her current interests are in the same field of study as her PhD thesis – Physical Layer Security techniques for the protection of wireless systems.

Dr. Dnyaneshwar S. Mantri is graduated in Electronics Engineering from Walchand Institute of Technology, Solapur (MS) India in 1992 and received Masters from Shivaji University in 2006. He has awarded PhD. in Wireless Communication at Center for TeleInFrastruktur (CTIF), Aalborg University, Denmark in March 2017. He has teaching experience of 25 years. From 1993 to 2006 he was working as a lecturer in different institutes [MCE Nilanga, MGM Nanded, and STB College of Engg. Tuljapur (MS) India].

From 2006 he is associated with Sinhgad Institute of Technology, Lonavala, Pune and presently working as Professor in Department of Electronics and Telecommunication Engineering. He is member of IEEE, Life Member of ISTE and IETE. He has written three books, published 15 Journal papers in indexed and reputed Journals (Springer, Elsevier, and IEEE etc.) and 23 papers in IEEE conferences. He is reviewer of international journals (Wireless Personal Communication, Springer, Elsevier, IEEE Transaction, Communication society, MDPI etc.) and conferences organized by IEEE. He worked as TPC member for various IEEE conferences and also organized IEEE conference GCWCN2014 and GCWCN2018. He worked on various committees at University and College. His research interests are in Adhoc Networks, Wireless Sensor Networks, Wireless Communications, VANET, Embedded Security specific focus on energy and bandwidth.

Ernestina Cianca is Assistant Professor at the Dept. of Electronic Engineering of the University of Rome Tor Vergata, where she teaches Digital Communications and ICT Infrastructure and Applications (WSN, Smart Grid, ITS etc.). She is the Director of the II Level Master in Engineering and International Space Law in Satellite systems for Communication, Navigation and Sensing. She is vice-director of the interdepartmental Center CTIF-Italy.

She has worked on wireless access technologies (CDMA, OFDM) and in particular in the waveforms design, optimization and performance analysis of radio interfaces both for terrestrial and satellite communications. An important part of her research has focused on the use of EHF bands (Q/V band, W band) for satellite communications and on the integration of satellite/terrestrial/HAP (High altitude Platforms) systems. Currently her main research interests are in the use of radio-frequency signals (opportunistic signals such as WiFi or specifically designed signals) for sensing purposes, and in particular device-free RF-based activity recognition/crowd counting/density estimation and localization; UWB radar imaging (i.e., stroke detection). She is author/co-author of around 100 papers in international journals and conferences.

Georgi Iliev received the MEng degree in Telecommunications from Technical University of Sofia (TUS), Bulgaria in 1990 and PhD degree in Adaptive Signal Processing from the same university in 1996.

He was with the Department of Telecommunications, Technical University of Sofia from 1993 as an Assistant Professor, and from 2003 as an Associate Professor. In 2011 he was elected as the Head of Department of Communications Networks and in 2012 became a full Professor in the same department. His interests are in signal processing, adaptive systems and algorithms, and noise cancellation. He has been working on systems for speech recognition in noisy environments, adaptive equalisation for communication channels, and the development of new computationally-efficient adaptive algorithms.

Homayoun Nikookar received his Ph.D. in Electrical Engineering from Delft University of Technology in 1995. He is an Associate Professor at the

Faculty of Military Sciences of the Netherlands Defence Academy. In the past he has led the Radio Advanced Technologies and Systems (RATS) research program, and supervised a team of researchers carrying out cutting-edge research in the field of advanced radio transmission. His areas of research include Wireless radio channel modeling, Ultra Wideband (UWB) Technology, MIMO, Multicarrier-OFDM transmission, Wavelet-based OFDM en Cognitive Radio Sensor Networks.

He has received several paper awards at international conferences and symposiums. Dr Nikookar has published 150 papers in the peer reviewed international technical journals and conferences, 15 book chapters and is author of two books: *Introduction to Ultra Wideband for Wireless Communications*, Springer, 2009 and *Wavelet Radio: Adaptive and Reconfigurable Wireless Systems based on Wavelets*, Cambridge University Press, 2013.

Mauro De Sanctis is Assistant Professor at the Department of Electronics Engineering, University of Roma "Tor Vergata" (Italy), teaching "Information Theory and Data Mining". From January 2006 to June 2008 he has been involved in the MAGNET Beyond European FP6 integrated project as scientific responsible of the activities on radio resource management. He is involved in the coordination of the Alphasat "Aldo Paraboni" Payload (Technology Demonstration Payload – TDP 5) scientific experiments for broadband satellite communications in Q/V band, jointly funded by ASI and ESA. He is serving as Associate Editor for the Signal Processing and Communication in Aerospace Systems area of the IEEE Aerospace and Electronic Systems Magazine and as Associate Editor for the Command, Control and Communications Systems area of the IEEE Transactions on Aerospace and Electronic Systems. His main areas of interest are: wireless terrestrial and satellite communication networks, data mining and information theory. He published more than 90 papers on journals and conference proceedings, 4 book chapters, one book and one patent.

Nicola Laurenti received his Laurea Degree in Electrical Engineering in 1995 and his Ph.D. in Electronic and Telecommunication Engineering in 1999 both from University of Padua, Italy. Since 2001 he has been an Assistant Professor at the Department of Information Engineering of University of Padua. In 2008–09 he was a Visiting Scholar at the Coordinated Science Laboratory of the University of Illinois at Urbana-Champaign, while in 1992–93 he was an exchange student in the Electrical Engineering and Computer Science Department at the University of California at Berkeley. He has authored 27 journal papers and 55 conference papers. His research interests mainly focus on wireless network security at lower layers (physical, data link and network), GNSS security, information theoretic security and quantum key distribution. He has been the coordinator of the activity by University of Padua in several projects related to the information security of satellite navigation systems, funded by the European Space Agency and the European Commission.

Peter Lindgren holds a full Professorship in Multi business model and Technology innovation at Aarhus University, Denmark – Business development and technology innovation and is Vice President of CTIF Global Capsule (CGC). He has researched and worked with network based high speed innovation since 2000. He has been head of Studies for Master in Engineering – Business Development and Technology at Aarhus University from 2014–2016. He has been researcher at Politechnico di

Milano in Italy (2002/03), Stanford University, USA (2010/11), University Tor Vergata, Italy (2016/2017) and has in the time period 2007–2011 been the founder and Center Manager of International Center for Innovation www.ici.aau.dk at Aalborg University, founder of the MBIT research group and lab – http://btech.au.dk/forskning/mbit/ – and is cofounder of CTIF Global Capsule – www.ctifglobalcapsule.com. He works today as researcher in many different multi business model and technology innovations projects and knowledge networks among others E100 – http://www.entovation.com/kleadmap/, Stanford University project Peace Innovation Lab http://captology.stanford.edu/projects/peace-innovation.html, The Nordic Women in business project – www.womeninbusiness.dk/, The Center for TeleInFrastruktur (CTIF) at Aalborg University www.ctif.aau.dk, EU FP7 project about "multi business model innovation in the clouds" – www.Neffics.eu, EU Kask project – www.Biogas2020.se. He is author to several articles and books about business model innovation in networks and Emerging Business Models. He has an entrepreneurial and interdisciplinary approach to research. His research interests are multi business model and technology innovation in interdisciplinary networks, multi business model typologies, sensing and persuasive business models.

Dr. Ramjee Prasad is a Professor of Future Technologies for Business Ecosystem Innovation (FT4BI) in the Department of Business Development, and Technology, Aarhus University, Herning, Denmark. He is the Founder President of the CTIF Global Capsule (CGC). He is also the Founder Chairman of the Global ICT Standardisation Forum for India, established in 2009. GISFI has the purpose of increasing of the collaboration between European, Indian, Japanese, North-American and other worldwide standardization activities in the area of Information and Communication Technology (ICT) and related application areas.

He has been honored by the University of Rome "Tor Vergata", Italy as a Distinguished Professor of the Department of Clinical Sciences and

Translational Medicine on March 15, 2016. He is Honorary Professor of University of CapeTown, South Africa, and University of KwaZulu-Natal, South Africa. He has received Ridderkorset af Dannebrogordenen (Knight of the Dannebrog) in 2010 from the Danish Queen for the internationalization of top-class telecommunication research and education.

He has received several international awards such as: IEEE Communications Society Wireless Communications Technical Committee Recognition Award in 2003 for making contribution in the field of "Personal, Wireless and Mobile Systems and Networks", Telenor's Research Award in 2005 for impressive merits, both academic and organizational within the field of wireless andpersonal communication, 2014 IEEE AESS Outstanding Organizational Leadership Award for: "Organizational Leadership in developing and globalizing the CTIF (Center for TeleInFrastruktur) Research Network", and so on. He has been Project Coordinator of several EC projects namely, MAGNET, MAGNET Beyond, eWALL and so on.

He is the founding editor-in-chief of the Springer International Journal on Wireless Personal Communications. He is member of the editorial board of several other renowned international journals and is the series editor of the Artech House Universal Personal Communications Series. Ramjee Prasad is a member of the Steering, Advisory, and Technical Program committees of many renowned annual international conferences, e.g., Wireless Personal Multimedia Communications Symposium (WPMC); Wireless VITAE, etc. He has published more than 30 books, 1000 plus journals and conferences publications, more than 15 patents, a sizeable amount of graduated PhD students (over 140) and an even larger number of graduated M.Sc. students (over 250). Several of his students are today worldwide telecommunication leaders themselves.

Sarmistha De Dutta, Member IEEE, has a B.Sc. (Physics) and B.E (Electronics Engineering) from Maharaja Sayajirao University, Baroda, India.

She earned her M.S. in Electrical Engineering from Columbia University, New York, USA. She currently works at NIKSUN, Inc. She previously worked at Bristol-Myers Squibb, Columbia University, and Box Hill Systems. Sarmistha is a member of the IEEE, its Women in Engineering (WIE), and the IEEE Professional Communication Society.

Shivprasad P. Patil received his B.E. degree in Electronics Engineering from University of Pune, India, in 1989 and Master degree from Swami Ramanad Tirth Marathwada University, Nanded, in 2000. He received his PhD from Aarhus University, Denmark in 2018. He is working as a Professor in the department of Information Technology in NBN Sinhgad School of Engineering, Pune, India. He has over 28 years of experience in academia as well as in industry. He has published 08 papers in international journals and conference proceedings in US, Europe and India. His research interests are in the areas of computer vision, multimedia data analysis and wireless multimedia communications.

Silvia Ceccato is a Ph.D. student at the University of Padua, Italy. She received her Master degree in Telecommunication Engineering from the same university in 2016 with a thesis on key management for GNSS access control and was awarded the "A Thesis for National Security" prize by the Italian government. Her research area is the security of GNSS, both at signal and data level, focusing on positioning resilience, key management architectures and data authentication and integrity protection. She won the Best Track Paper award at ION PLANS Symposium in 2018, as first author. She has participated in several activities in cooperation with the European Space Agency on authentication and integrity protection for the Galileo Open Service. She has also been working on visual aided spoofing detection visiting the University of Queensland in Brisbane, Australia.

Simone Di Domenico received both Bachelor and Master degrees in Internet technology engineering from the University of Roma "Tor Vergata", in 2012 and 2014, respectively. He got the Ph.D. degree in Electronic Engineering at the University of Roma "Tor Vergata" in 2018. Currently, he is a postdoctoral researcher at the University of Roma "Tor Vergata" and his main research interests include the RF device-free human activity recognition and the RF device-free people counting.

Sofoklis A. Kyriazakos graduated in 1993 a Greek-American school, namely Athens College and obtained his Master's degree in Electrical Engineering and Telecommunications in RWTH Aachen, Germany in 1999. Then he moved to the National Technical University of Athens, where he obtained his Ph.D. in Telecommunications in 2003. He also received an MBA degree in Techno-economic systems from the same University. Sofoklis holds the position of Associate Professor in the University of Aarhus, where his activities are focused around Innovation Management and Entrepreneurship. He has managed, both as technical manager and coordinator, a large number of multi-million ICT projects, both at R&D and industrial level. Sofoklis has published more than 100 publications in international conferences, journals, books and standardization bodies and has more than 400 citations. In 2006 Sofoklis founded Converge SA, a startup with specialization in software integration. In 2016 after having steered successfully the company from the position of Managing Director, he resigned to co-found Innovation Sprint Sprl and he is currently the CEO of the company. Sofoklis is member of Board of Directors in 2 companies in Greece and Canada and has also been a member of the BoD of Athens Information Technology, a Center of Excellence for Research and Education.

Tommaso Rossi is an Assistant Professor of Telecommunications Engineering at the University of Rome Tor Vergata. (teaching Digital Signal Processing, Multimedia Processing and Communication and Signals). His research activity is focused on Space Systems, Extremely High Frequency Satellite and Terrestrial Telecommunications, Satellite and Inertial Navigation Systems, Digital Signal Processing for Radar and TLC applications. He is currently Co-Investigator of the Italian Space Agency Q/V-band satellite communication experimental campaign carried out through the Alphasat Aldo Paraboni payload. He is Associate Editor for the Space Systems area of the IEEE Transactions on Aerospace and Electronic Systems.

Vandana Milind Rohokale received her B.E. degree in Electronics Engineering in 1997 from Pune University, Maharashtra, India. She received her Masters degree in Electronics in 2007 from Shivaji University, Kolhapur, Maharashtra, India. She has received her PhD degree in Wireless Communication in 2013 from CTIF, Aalborg University, Denmark. She is presently working as Professor in Sinhgad Institute of Technology and Science, Pune, Maharashtra, India. Her teaching experience is around 22 years. She has published one book of international publication. She has published around 35 papers in various international journals and conferences. Her research interests include Cooperative Wireless Communications, AdHoc and Cognitive Networks, Physical Layer Security, Digital Signal Processing, Information Theoretic security and its Applications, Cyber Security, etc.

Viktor Stoynov received his BSc and MSc degrees in Telecommunications from the Technical University of Sofia (TUS), Bulgaria, in 2012 and 2014 respectively. During his PhD, from December 2015 to April 2016 he took part in the Erasmus educational program in the Technical University – Vienna, Austria, where he contributed to the development of the Vienna LTE-A simulators. In 2018 he received his PhD degree in Theoretical Fundamentals of Telecommunications at the TUS on the topic "Investigation of Resource Allocation in Abstract Modelled Indoor Wireless Communication Environments in New Generation Networks". Currently he is Senior Assistant Professor in the Department of Communication Networks, Faculty of

Telecommunications, TUS. His current interests are in the field of Low Power Wide Area Networks, interference management in heterogeneous networks and intelligent resource allocation strategies. Currently, Senior Assist. Prof. Stoynov has 14 papers and publications.

Vladimir Poulkov, has received his MSc and PhD degrees at the Technical University of Sofia, Bulgaria. He has more than 30 years of teaching, research and industrial experience in the field of telecommunications. He is senior IEEE Member with expertise in the field of information transmission theory, modulation and coding, interference suppression, resource management for next generation telecommunications networks, cyber physical systems. Currently he is Head of "Teleinfrastructure" and "Electromagnetic Compatibility of Communication Systems" R&D Laboratories at the Technical University of Sofia, Bulgaria. He is Chairman of Bulgarian Cluster Telecommunications, Vice-Chairman of the European Telecommunicaitons Standartization Institute (ETSI) General Assembly.

Yapeng Wang obtained Master Degree from Beijing University of Post and Telecommunication in 2008. Currently she is a guest researcher in CTIF in the field of Network Neutrality. Before 2008, she worked in Teleinfor Institute of

CATR as a researcher and the field is telecommunication regulation, policy and market. Till now, worked in International Cooperation Department of CATR, and responsible for EU projects and ITU-D issues in China. The projects include:

- 2011–2012 Promotion of Green Economic Growth by Broadband Network
- 2012–2013 Open China ICT Project – Observation of the Chinese telecom' development
- 2012–2013 Implementing and planning outline of 'Smart Qianhai's policy (Qianhai is a region of Shenzhen)
- 2012–2014 International standard assessment of China Unicom from 2010 to 2010
- 2014 WTDC: China Reception, Editorial Committee, Election for Mr. ZHAO Houlin
- 2014 PP: China Reception, Editorial Committee, Election for Mr. ZHAO Houlin
- 2014 Application for ITU Centers of Excellence (CoE) in Conformance and Interoperability (C&I)
- 2014 Application for Conformance and Interoperability Test lab

Zlatka Valkova-Jarvis received her MSc degree in Electrical Engineering and PhD degree in Theoretical Fundamentals of Telecommunications at the Technical University of Sofia (TUS), Bulgaria. Currently she is Associate Professor in the Department of Communication Networks, Faculty of Telecommunications, TUS and is the Vice-Dean of this Faculty. Assoc. Prof. Jarvis's academic experience includes teaching BSc courses in Communication Circuits and Digital Signal Processing (taught in both Bulgarian and English), and MSc courses in Digital Signal Processing in Telecommunications. Her main research interests include Digital Signal Processing,

Communication Circuits, and Efficient Digital Filtering. In these research areas she has over 60 papers and publications, and is author and co-author of over 10 textbooks, manuals and book-chapters. She has participated in various research projects (nationally- and internationally-funded) in different areas of Telecommunications theory and practice, and has been a visiting researcher/lecturer in a number of European universities. Assoc. Prof. Jarvis is a member of the Bulgarian sections of the IEEE Signal Processing and Circuits and Systems.

About the Editors

Dr. Ramjee Prasad is a Professor of Future Technologies for Business Ecosystem Innovation (FT4BI) in the Department of Business Development, and Technology, Aarhus University, Herning, Denmark. He is the Founder President of the CTIF Global Capsule (CGC). He is also the Founder Chairman of the Global ICT Standardisation Forum for India, established in 2009. GISFI has the purpose of increasing of the collaboration between European, Indian, Japanese, North-American and other worldwide standardization activities in the area of Information and Communication Technology (ICT) and related application areas.

He has been honored by the University of Rome "Tor Vergata", Italy as a Distinguished Professor of the Department of Clinical Sciences and Translational Medicine on March 15, 2016. He is Honorary Professor of University of CapeTown, South Africa, and University of KwaZulu-Natal, South Africa. He has received Ridderkorset af Dannebrogordenen (Knight of the Dannebrog) in 2010 from the Danish Queen for the internationalization of top-class telecommunication research and education.

He has received several international awards such as: IEEE Communications Society Wireless Communications Technical Committee Recognition Award in 2003 for making contribution in the field of "Personal, Wireless and Mobile Systems and Networks", Telenor's Research Award in 2005 for impressive merits, both academic and organizational within the field of wireless andpersonal communication, 2014. IEEE AESS Outstanding Organizational Leadership Award for: "Organizational Leadership in developing and globalizing the CTIF (Center for TeleInFrastruktur) Research Network", and so on. He has been Project Coordinator of several EC projects namely, MAGNET, MAGNET Beyond, eWALL and so on.

He is the founding editor-in-chief of the Springer International Journal on Wireless Personal Communications. He is member of the editorial board of several other renowned international journals and is the series editor of the Artech House Universal Personal Communications Series. Ramjee Prasad is a member of the Steering, Advisory, and Technical Program committees of many renowned annual international conferences, e.g., Wireless Personal Multimedia Communications Symposium (WPMC); Wireless VITAE, etc. He has published more than 30 books, 1000 plus journals and conferences publications, more than 15 patents, a sizeable amount of graduated PhD students (over 140) and an even larger number of graduated M.Sc. students (over 250). Several of his students are today worldwide telecommunication leaders themselves.

Leo P. Ligthart was born in Rotterdam, the Netherlands, on September 15, 1946. He received an Engineer's degree (cum laude) and a Doctor of Technology degree from Delft University of Technology. He is Fellow of IET and IEEE and Academician of the Russian Academy of Transport.

He received Honorary Doctorates at MSTUCA in Moscow, Tomsk State University and MTA Romania.

Since 1988, he held a chair on Microwave transmission, remote sensing, radar and positioning and navigation at Delft University of Technology. He supervised over 50 PhD's.

He founded the International Research Centre for Telecommunications and Radar (IRCTR) at Delft University.

He is founding member of the EuMA, chaired the first EuMW in 1998 and initiated the EuRAD conference in 2004. Member BoG of IEEE-AESS from 2013–2019.

Currently he is emeritus professor of Delft University, guest professor at Universities in Indonesia and China, Chairman of CONASENSE.

His areas of specialization include antennas and propagation, radar and remote sensing, satellite, mobile and radio communications. He gives various courses on radar, remote sensing and antennas. He has a H-index of 37, he published over 700 papers, various book chapters and 7 books.